The Philosophical Foundations of Bayesianism

Bayesianism is the view that rational degrees of belief should follow probability theory. In this book, philosopher Kenny Easwaran defends Bayesianism as an effective normative framework when applied to all "doxastic subjects": humans, animals, artificial intelligences, groups, and even hypothetical aliens or angels.

The book addresses major challenges: concerns about numerical precision and computational tractability, problems with infinitesimals and regularity, difficulties with act-state dependence, and debates over evidence. Easwaran employs measurement theory to show probabilistic representation is analogous to representing temperature – different scales (Fahrenheit/Celsius, probability/odds) can represent the same underlying non-numerical reality.

Later chapters examine how beliefs should update through conditionalization, while distinguishing different sources of rational norms. Easwaran defends a notably permissive view: any probability function is internally rational, as reason alone cannot dictate privileged priors. However, "ecological rationality" explains why certain belief systems perform better in specific environments, acknowledging both Bayesianism's abstract permissiveness and practical constraints making certain credal states more successful.

Kenny Easwaran is Professor of Logic and Philosophy of Science at the University of California, Irvine. He works on many issues in formal epistemology, decision theory, and the philosophy of mathematics, as well as having interests in how these topics relate to collective agency, artificial intelligence, games and puzzles, and urban planning.

The Philosophical Foundations of Bayesianism

An Opinionated Introduction

Kenny Easwaran

Routledge
Taylor & Francis Group
NEW YORK AND LONDON

Designed cover image: Getty Images

First published 2026
by Routledge
605 Third Avenue, New York, NY 10158

and by Routledge
4 Park Square, Milton Park, Abingdon, Oxon, OX14 4RN

Routledge is an imprint of the Taylor & Francis Group, an informa business

ISBN: 978-1-138-64770-1 (hbk)
ISBN: 978-1-138-64771-8 (pbk)
ISBN: 978-1-315-62691-8 (ebk)

DOI: 10.4324/9781315626918

Typeset in Latin Modern font
by KnowledgeWorks Global Ltd.

Contents

CHAPTER 1

Belief comes in degrees

1.1 INTRODUCTION

1.1.1 Bayesianism in epistemology

Epistemology is the study of knowledge and related mental states. Many historical philosophers such as Plato and Descartes had a conception of knowledge that brought with it some strong kind of certainty, to the point that it is an open question for them whether anyone really knows anything. But since at least Malcolm (1952) and Gettier (1963), there has been more of a focus on a conception of knowledge that doesn't require infallible certainty. Although Gettier argued that knowledge is more than just belief that is both justified and true, the study of belief and what it takes for belief to be justified have been central topics in epistemology for the past 80 years. While "epistemology" is still the label used for the field broadly, many epistemologists distinguish "epistemic" issues, involving knowledge itself, from "doxastic" ones, involving belief more generally.

Although most work in epistemology treats it from the perspective of the individual, in recent decades there has been a growth of work in social epistemology. Some of this includes work on how one individual's knowledge, belief, and justification are affected by those of others (for instance, Coady (1990) and Lackey (2008) on the role of testimony or Kelly (2005) and Feldman (2006) on the role of disagreement). Some is about what it takes for a *group* of people to have knowledge, belief, or justification (for instance, Gilbert (1987), List and Pettit (2011) and Tollefsen (2015)). Some is on other epistemologically relevant features of groups (for instance, Anderson (2006) and Landemore (2011) on the

DOI: 10.4324/9781315626918-1

epistemology of democracy, Frost-Arnold (2016) on social media, and Nguyen (2020) on echo chambers).

Formal epistemology is the branch of epistemology that applies mathematical methods to both epistemic and doxastic issues. There are a great number of different projects that apply different sorts of mathematical methods to different aspects of epistemology. There are epistemic logics that develop a formal logic with an operator that represents knowledge (Rendsvig et al., 2023). There are logics of belief revision, which develop systematic methods for revising one's beliefs, while remaining logically consistent, even when learning new evidence that contradicts some of what one previously believed (Alchourròn et al., 1985). There are various agent-based computer simulations of the flow of information in social networks (Zollman, 2007 and Weatherall et al., 2020). But the biggest amount of work in formal epistemology has focused on the study of uncertainty, and conceptions of belief that come in degrees.

While there are many formal theories of uncertainty and degrees of belief (see for example the various theories described by Walley (1991) and Halpern (2017)), the one that is most well-developed is Bayesianism. The core of Bayesianism is the use of the mathematical theory of probability as a formal model of degree of belief. (The basic outlines of the theory of probability will be given in Section 1.4.) The term "Bayesian" itself is used because of a paper published by the Rev. Thomas Bayes (1763), where several important early results in probability were first stated and proved. (There is debate about whether it is appropriate to attribute anything like the modern Bayesian view to Bayes himself—for more on this historical question, see Chapter 1 of Earman (1992).)

1.1.2 Bayesianism in different academic disciplines

Bayesianism has become a major topic in many different academic disciplines. Because these disciplines approach Bayesianism from different directions, they tend to focus on different aspects of it and contrast it with different alternatives. Thus, people from one discipline who consider themselves opposed to Bayesianism might have a viewpoint that people from another discipline might still think of as Bayesian, and vice versa.

Within the field of statistics, probability and its role in interpreting scientific observations and experiments is central. The Bayesian view of probability, on which it represents the degree of belief of some (actual or hypothetical) reasoner, is contrasted with a frequentist account of probability, on which it merely measures the fraction of outcomes of some particular type in a repeatable scenario. For non-repeatable things (like a hypothesis about a law of nature that has either always been true or never been true, or a hypothesis about some historical event that occurred only once, like the cause of the extinction of the dinosaurs, or the true number of ballots for each candidate in the 2000 US presidential election), there either is no frequentist probability, or it is trivially 0 or 1. Bayesian statisticians are happy to assign prior probabilities representing some reasoner's degrees of belief in these hypotheses, and show how various observations should update these probabilities. Frequentists object to the subjectivity involved in using some particular reasoner, and instead aim to describe repeatable experimental and observational methods that anyone can use in ways that ensure only a small fraction of people will make incorrect judgments about these hypotheses (assuming the theoretician is using the right probability model for the application). But even Bayesian statisticians usually aim to develop prior probabilities that might be considered sufficiently objective for scientific practice. Some prominent Bayesian statisticians (like Gelman and Shalizi (2013)) deny that their work should be used in ways that a philosopher would identify as "Bayesian".

In the fields of economics and psychology, entirely separately from the role of statistics in helping theorists interpret empirical evidence, Bayesian probability forms a part of the subject matter of the field. Both economists and psychologists are interested in the behavior of individuals who are acting under uncertainty. Bayesianism provides one such model of how individuals might behave. In some contexts, the assumption that people behave like Bayesian subjects allows economists and psychologists to infer people's degrees of belief by observing their actions, or conversely to predict their actions by understanding their degrees of belief. However, several important lines of work showed that actual people fail to follow the predictions of Bayesian theory, and developed alternate theories that seem to match human behavior better, like the "prospect theory" of Kahneman and Tversky (1979). In these fields, one often associates "Bayesianism" with the claim that actual human departures from idealized Bayesian theory is not too large, while "non-Bayesianism" is the claim that actual human departures are so substantial that a different descriptive framework is needed.

But economists and psychologists also engage with the question that I think of as most distinctively philosophical about Bayesianism, and which this book will focus on. This is the question not of how humans *do* behave in the world, but rather how we *should* behave. Philosophers distinguish these as the "descriptive" and "normative" question (though economists and some others tend to use the word "positive" where philosophers use "descriptive"). Kahneman himself stuck to the idea that Bayesianism is the correct normative theory of behavior under uncertainty, and treated prospect theory as a descriptive theory, though some psychologists and philosophers have attempted to develop competing normative theories. (Note that the kind of normativity being appealed to here is taken to be a distinctively "epistemic" kind of normativity. The primary question is not about how one *morally* should behave or *legally* should behave, but something more like how one *rationally* should behave. Refer to Basu (2023a,b) for more on how moral normativity might be relevant for epistemology.)

1.1.3 Outline

This book is aimed at developing the philosophical foundations for Bayesianism as a normative theory of doxastic states. In particular, the claim is that Bayesianism is the theory of doxastic states that are optimal at fulfilling their function, which is either to guide behavior that is successful from the subject's point of view or to represent the world accurately. The rest of this chapter gives the basic background for this—Section 1.2 discusses what makes a state "doxastic", and what sort of creature has doxastic states; Section 1.3 considers what doxastic states are about; Section 1.4 develops the basic theory of the "degree of belief" attitude that is central to the doxastic state, and gives the basics of what it means for it to satisfy the axioms of probability; and Section 1.5 discusses how this "degree of belief" concept relates to the ordinary concept of "belief". Chapters 2 and 3 then discuss two different ways to ground the normativity of doxastic states, either in connection to preference and action or in connection to the truth. Chapters 4 and 5 discuss the mathematical aspects of Bayesianism, addressing concerns about what it means for doxastic states to be measured numerically, and about what a particular number would mean. Chapter 6 turns to the question of how doxastic states should evolve over time, describing the orthodox Bayesian theory of updating and showing how it follows from either grounding of the normativity described in Chapters 2 and 3. Chapters 7 and 8 return to the over-arching normative questions, with Chapter 7 arguing that several aspects of Bayesian normativity actually arise from different sources, and Chapter 8 addressing some big questions for rational normativity generally.

One general thing to note—although Bayesianism is a very mathematical theory, this book is focused more on the conceptual aspects of it than the mathematical aspects. Some mathematical tools will inevitably be used throughout the book, but none of this will be very extensive. This book does not include any extensive training in the use of mathematics within Bayesianism. Furthermore, while this book touches on many general issues in epistemology to give a philosophical foundation for Bayesianism, it doesn't aim to *use* Bayesian methods to address other specific questions in epistemology. For readers interested in practice with the mathematics of Bayesianism and its epistemological applications, books like Bradley (2015) and Titelbaum (2022a,b) will be more relevant. This book just focuses on the philosophical foundations of Bayesianism.

1.2 METHODOLOGY

1.2.1 The word "belief"

Bayesianism is a theory of doxastic states. But what are doxastic states? The first approach we might consider is to look at the word "belief" that is often used to define it. First, I'll note that "belief" and "believe" are words that philosophers use somewhat more broadly than many other people. Many people use the word "believe" primarily for more spiritual or religious contexts, but philosophers use the word in contexts where ordinary people might more naturally use "think". But philosophers aren't unique here—both uses exist in ordinary language, as W.C. Fields noted when saying "Everybody's got to believe in something. I believe I'll have another beer". When philosophers talk about doxastic states, we are interested in states that are like believing or thinking something is true.

Belief in this sense is a kind of mental state. But there are lots of other mental states, which have more or less to do with beliefs. We have knowledge, perceptions, plans, imaginations, intentions, suspicions, thoughts, emotions, habits, desires, wishes, and many other things. We could try to just come up with some categories to divide these. Some of these, like knowledge and perception, might be classified as "epistemic" (from the Greek word "$\epsilon\pi\iota\sigma\tau\eta\mu\eta$" for "knowledge"), and others, like desires and wishes, might be classified as "bouletic" (from the Greek word "$\beta o\upsilon\lambda\eta$" for "wish" or "want"), while we identify suspicions, beliefs, and thoughts as "doxastic" (from the Greek word "$\delta o \xi \alpha$" for "belief"). But should imaginations or intentions get categorized as doxastic? And how many categories should there be?

1.2.2 Conceptual analysis

This is a question of conceptual analysis, which has been a core project in philosophy. I see two broad families of methods that philosophers have used to approach the project of conceptual analysis. One of these families, which we might call "analytic", presumes that we have already identified a concept, and tries to define it by identifying its essence. The other family, which we might call "pragmatic", looks at what we want a concept to *do*, and tries to identify what features are relevant for doing that thing, which may or may not coincide with some concept we are already using. Analytic approaches treat a conceptual analysis as correct or incorrect, while pragmatic approaches treat it as more or less useful.

There are various methods I count as being broadly "analytic" in this sense. The "Canberra Plan" starts by identifying a consistent set of basic "platitudes" about a concept, and then says that whatever satisfies those platitudes is the extension of that concept. This method is one that David Lewis and Frank Jackson made explicit in the 1990s, based on an idea Lewis (1970) derives from the work of Frank Ramsey (though Ramsey himself didn't work this way). The method of "reflective equilibrium", introduced by John Rawls in the 1970s (Knight, 2023), is a bit more flexible—it starts with intuitions about the concept *and* intuitive judgments about cases, and then modifies each in light of the other, until a definition is arrived at that aligns with as many of each as possible. The "ordinary language" philosophy of Oxford in the 1950s and 1960s is similar, whose practitioners paid close attention to the details of ordinary use of a word and tried to identify the meaning with that detailed use. One might even see Kant's "constitutive principles" for rationality as doing some of this same work, taking the idea of a thinking being for granted, identifying what features are necessary for thinking, and then concluding that those features must hold of us.

"Pragmatist" methods have also arisen at various points. In the 1940s, Rudolf Carnap described a method of "explication" (see Appendix D on "Methodology", of Leitgeb and Carus (2020)) by which one replaces a vague natural language concept and creates one or more precise definitions that come close to it, and sees which do good jobs of whatever you wanted to use the concept for. Carnap gets this method from mathematics, where a concept like "dimension" gets explicated in various different ways that are more or less effective for linear spaces or manifolds or fractals. More recently, Haslanger (2012) describes an "ameliorative" method of conceptual analysis, where she aims to replace ordinary language concepts with ones that make society better. Other philosophers have engaged in a related project of "conceptual engineering" (Burgess et al., 2020) where we try to design new concepts that will do useful work within philosophy.

My method for understanding "doxastic states" will be of the more pragmatic sort. I'll identify some things that we often want to think about when thinking about doxastic states, and use them to build definitions. There will be two distinct ways of thinking about doxastic states that I consider, but an interesting fact is that both give rise to basically the same sort of Bayesian normativity. It may thus be useful to think of these as two different explications of the concept of "doxastic state", but many applications of Bayesianism won't depend on which is used. I'll treat them as two parts of one disjunctive explication, but there may be other ways to understand these ideas. There may be other explications one can come up with, and it will be an interesting question whether they also provide a foundation for Bayesianism.

1.2.3 Functionalism

My account will be pragmatist in another way as well. Both of the explications I give of the concept of "doxastic state" will be in terms of what the state is used for by the creature that has it. That is, I won't try to define belief by saying what belief *is*, but rather what belief is *for*. This way of approaching epistemology by starting with the roles for belief and then investigating appropriate norms for these roles fits well with functionalism in the philosophy of mind.

Functionalism is the view that what makes something mental is not being constituted out of special "mind-stuff" (or even being constituted out of neurons organized in a certain way), but rather is just playing a particular role in a system. Whether human minds are immaterial entities that control the body through a spiritual connection, or collections of neurons that control the body through chemical interactions with muscles, or something else, if it is possible for other physical beings manifested in other physical ways to control their bodies in some similar enough way, then these physical beings would count as having minds as well (Levin, 2013).

One advantage of this functionalist account of doxastic states for my project is that it gives a direct approach to normativity. As David Hume observed, it's hard to derive an ought from an is—if we just say that a bridge is a solid structure that crosses a valley or body of water, or that a knife is a sharp object with a handle, it's hard to say what a bridge ought to be like, or what a knife ought to be like. This definition doesn't help us figure out whether they should be made of steel or lace or obsidian or plastic. But if we say that a bridge is *for* people or vehicles to use to cross a valley or body of water, or that a knife is *for* cutting food or *for* use as a weapon, then we can see why it might be better to make them out of steel than of lace, and can figure out whether obsidian or plastic might be a better choice for the particular one we are making.

Thus, my pragmatic analysis of the concept of "doxastic state" will focus on the uses a doxastic state has for the creature that has it.

1.2.4 Pragmatic and alethic roles for the doxastic state

The two functional roles that I will use as independent characterizations of a "doxastic state" are ones that I will call "pragmatic" and "alethic".

The pragmatic role for a doxastic state is the role it plays in governing behavior. ("$\pi\rho\alpha\gamma\mu\alpha$" is Greek for "deed".) David Hume's theory of motivation suggests that a person's intentional actions are guided by their beliefs and desires. Roughly, someone will do action A if they *believe* that A will best satisfy their *desires*. Beliefs about one part of the world, together with desires about a related part, will result in action. Chapter 2 develops this account more precisely and shows how it leads to Bayesianism.

The alethic role for a doxastic state is the role of accurately representing the world. ("$\alpha\lambda\eta\theta\epsilon\iota\alpha$" is Greek for "truth".) Some epistemologists (see Christensen (1991, 1996, 2001), Joyce (1998), and Eriksson and Hájek (2007)) think that we should be able to conceive of epistemology as applying even to creatures that purely observe, rather than having any desire to act in the world. They think that for such a creature, there is still a sense in which its beliefs are better if they are more accurate and worse if they are less accurate. Chapter 3 develops this account more precisely and shows how it also leads to Bayesianism.

1.2.5 Three levels of pragmatism

I have used the word "pragmatic" at three different levels. The first was at the meta-philosophical level, in saying how we should go about defining concepts, by seeing what use the *concept* has for us *as philosophers.* It was contrasted with an "analytic" approach on which the concept is taken as given. The second was at the philosophical level, saying that it is useful for philosophers to understand the normativity of doxastic states by seeing what use the *states* have *for the creatures that have them.* It was contrasted with an approach where we try to describe the intrinsic nature of a doxastic state. The third is at the object level, by focusing on the use of the doxastic state for governing action. It is contrasted with an "alethic" account, focusing on the role of accuracy, independent of action. As you might suspect, I favor the pragmatic role at this third level as well but will discuss the alethic role too, since each can independently give a foundation for Bayesianism.

1.2.6 Direction of fit

Both the pragmatic and alethic roles for doxastic states like belief and degree of belief give them an interesting connection to bouletic states like desire and preference. On the pragmatic account, it is the doxastic and bouletic states that together control behavior, but something must be said about the difference in how they work together. On the alethic account, both doxastic and bouletic states share an aim of accurately representing the world, though this may not be obvious immediately.

This connection was investigated by Anscombe (1963), and the difference between the two has come to be described in terms of "direction of fit". She illustrates this with an example. Imagine that there is a shopper who enters a store with a grocery list, and finds the items in the store and puts them in the cart. Imagine that they are being watched by a detective, who sees the items the shopper is putting in their cart and writes them down on a list. If the shopper and the detective have both done what they are aiming to do, both lists will match the items in the cart. If there is a mismatch between the cart and the list, something has gone wrong.

But fixing this wrong is different for the shopper and for the detective. For the shopper's list to do its job right, the shopper should replace the items in the cart until they match the words on the list. But for the detective's list to do *its* job right, the detective should replace the words on the *list* until they match the items in the *cart.* The shopper's list is thus said to have a "world-to-mind" direction of fit, which is characteristic of bouletic states—the cart is being shaped to match the list; the world is being shaped to match the mind. The detective's list is said to have a "mind-to-world" direction of fit, which is characteristic of doxastic states—the list is being shaped to match the cart; the mind is being shaped to match the world.

1.2.7 Humans, and the "five A's"

Both of these functionalist accounts let us say what doxastic states should be like in order to do their jobs well. But they also let us say what kinds of creatures can have doxastic states. The obvious case that most epistemology has focused on is humans. But it's worth noting that humans aren't the only creatures that have pragmatic or alethic goals. Keeping the others in mind will help us focus on the broad functional role of doxastic states, rather than the empirical idiosyncrasies of humans. I have identified five classes of creatures that I think are worth keeping in mind, and some helpful names that all begin with the letter "A" as a helpful mnemonic. I will refer to them as "the five A's" when I feel it's important to remind the reader to think beyond humans:

- Animals—Most people who observe dogs and ravens and octopuses identify them as having practical goals, and also as having representations of the world that could be more or less accurate. I use the more general term "Animal". However, this category stands in for *any* natural biological life on Earth that has some sort of states that either help guide their behavior, or aim at accurately representing the world. After watching sped-up videos of plant tendrils exploring their surroundings for a place to securely attach before thickening into a vine, or of roots growing into a medium looking for water and nutrients, and seeing similar behaviors from fungi and bacteria (as well as complex signaling systems they have with each other), I suspect that doxastic states of the sort I am interested in extend well beyond the animal kingdom. (Figdor, 2018)

- Aliens—We don't know if there is life elsewhere in the universe, but many of us suspect that there is, given how large the universe is. This life would likely be unlike Earth life in many ways, but if it has goals it wants to achieve through behavior, or thinks about the world around it in ways that might be more or less accurate, it will have doxastic states in the senses I am interested in as well. This category stands in for any sort of natural life with origins outside of Earth that can usefully be thought of as having doxastic states.

- Artificial Intelligence—In addition to natural biological life, there has been a project over the past seven decades to create artificial systems that think about the world around them. For decades, there have been various databases fed by humans, symbolically-structured programs controlling sensors, and other intelligently-designed systems. But in the past decade and a half, these have been joined by neural systems of surprising power, from the AlexNet image classifying system, to the increasingly-multimodal Large Language Models built by OpenAI, Anthropic, Google/DeepMind, and other companies, to the self-driving systems in automobiles designed by Waymo and Tesla. Whether or not these systems are truly "intelligent" or have minds, many of them have states that aim to govern behavior toward some sort of goals, and others have states that aim to represent the world accurately even if the system itself doesn't use them in behavior. Thus, these systems have doxastic states in the sense I am discussing. The normative theory of Bayesianism applies to them, and they can provide useful examples for thinking about epistemology.

- Associations—Here the "A" is a bit contrived. It is more natural to call these "groups". We often speak as though companies, governments, and teams have goals they are trying to achieve, and have representations of the world that can be more or less accurate. To the extent that this is a useful way to think of them, they have doxastic states in the general sense that I am interested in. This category stands in for any system that is useful to think of as having doxastic states, that is itself composed of other systems that are themselves useful to think of as having doxastic states. Importantly, although there is a major debate about whether groups can have minds, I can follow Silver (2022) and separate the questions that are relevant for my concept of doxastic state from questions about "minds". What matters is whether these groups have states that govern behavior in the appropriate way, or aim at accuracy.

- Angels—Members of many religious traditions believe that there are supernatural entities that act in the world, and have beliefs about the world. I don't personally believe any such beings actually exist. But it is still sometimes helpful to imagine them, and imagine how they would function, to understand which aspects of a doxastic state are independent of any physical manifestation. This category stands in for any non-natural, non-artificial creature that might usefully be thought of as having doxastic states, whether angels or demons, ghosts or deities, spirits or anything else.

These categories (and humans) aren't fully exclusive of each other. Most modern people believe that humans are Animals. But among those that don't, they often follow Descartes (1641) and identify humans as immaterial souls that only contingently interact with the material world through their bodies—on my account, this would make us Angels. In trying to define "machines", Turing (1950) notes that there is a sense in which humans born in the usual way are "artificial" in the general sense of being made by humans, so we might count ourselves among the Artificial Intelligences. I also think that it's useful to think of Associations as being a kind of Artificial Intelligence, just one that is made of humans rather than being made out of silicon chips (or quantum computers, or clockworks, or any of the other ways that artificial intelligences have been imagined). Corporations, governments, and systems like Wikipedia are all artificial means by which people have tried to create entities with greater power to act or to represent the world than that of individual humans. (For a more detailed analysis of this sort, refer to Danzig (2022).) Conversely, many Artificial Intelligences are designed in ways that consist of multiple interacting parts, each of which is itself a kind of Artificial Intelligence. In Chapters 7 and 8, I will return to this theme and suggest that it may be useful to think of *all* creatures with doxastic states on the model of Associations, such that social epistemology is much more central to epistemology than has usually been thought. Because of this diversity of creatures that might be able to have doxastic states, I will use the term "doxastic subject" to neutrally refer to whatever sort of creature is involved.

1.2.8 Normative, not empirical

I will sometimes point to examples of one or another sort of actually existing doxastic subject to illustrate some of the conceptual relationships I am discussing. It may well be useful to look at them all more thoroughly to get an empirical understanding of some of the variety of features that doxastic states might have. But we should be careful to note that the mere fact that an actual doxastic subject exhibits a certain feature is merely suggestive that this feature is a good one. Actual subjects are usually pretty good at what they are doing, but do depart from the ideal in many ways as well.

One is just the obvious fact that not everything that has a function is ideal at carrying it out. Carrying out a function often requires resources of time, energy, and computational power that may not always be available. (For a useful approach to questions of rationality given resource limitations, see Icard (2026).) Furthermore, actual subjects often malfunction in various sorts of ways—a government might violate its constitution; a person might have long-term mental health issues, or short-term issues like distraction or sleep; an artificial intelligence might be misaligned. Furthermore, artificial intelligences and groups might just happen to be designed in ways that are less than ideal. Animals (including humans) with long evolutionary histories in a particular environment may often be close to ideal for the kinds of questions they face in that environment—but may have been transplanted into other environments that their doxastic states are not optimized for.

But another issue to keep in mind is that actually existing doxastic subjects have *many* goals they are trying to achieve, not just their doxastic functions. Animals aim to reproduce, governments aim to empower the populace, and artificial intelligences aim to promote outcomes aligned with human values. As I will discuss in greater depth throughout the book, having states that effectively fill the pragmatic function or the alethic function of a doxastic state will *usually* help subjects achieve these other goals. But not always.

Some empirical studies of humans have identified ways that human belief in fact serves functions somewhat different from the doxastic roles we have identified. Although belief usually governs action, Schwitzgebel (2001) and Gendler (2008) point to cases where only some of the practical functions of belief are played. A person who consciously knows the glass ledge on a skyscraper's observation deck is secure may be able to use this knowledge to close her eyes and walk forward, even while she is unable to walk to the edge with her eyes open. A person who is consciously anti-racist may be able to use their belief in racial equality to set up hiring processes for their company, even if, due to subconscious biases, they are personally unable to use this belief to equitably evaluate interview performances. Van Leeuwen and Lombrozo (2023) and Vesga et al. (2025) argue that some beliefs (particularly those about religion and spirituality) serve additional functions (more connected to social signaling and group formation) beyond the functions of governing action or accurate representation. Empirical studies of other doxastic subjects may well reveal more functional roles that exist.

But these two types of ways that actual doxastic subjects depart from ideal may sometimes cancel out. A lack of resources or a malfunction may sometimes detract from the doxastic function of a state—or may sometimes detract from the non-doxastic function while promoting the doxastic function. Identifying whether children or people with depression or misaligned AIs or non-democratic governments or injured animals turn out to be more or less doxastically rational than other subjects of their type is a tricky empirical question that goes beyond the scope of this book. While I will sometimes point to actual features of actual subjects to illustrate some concepts, the normative work I focus on will all involve direct conceptual connections to the identified function of a doxastic state, whether pragmatic or alethic.

1.3 WHAT ARE THE OBJECTS OF BELIEF?

1.3.1 Representation

With the idea that the doxastic state has a mind-to-world direction of fit (just as the bouletic state has a world-to-mind direction of fit), it must in some sense "represent" the world. This makes it different from mental states like some moods or emotions—one can be nervous or excited without there being anything in particular that this state is about. But this doesn't yet say what it means to represent, or how it manages to represent.

I won't say much about how it is possible to represent. For humans and other animals, I suspect that something like the teleosemantic project can help explain how evolution provided states whose function is to represent the world (Millikan, 1984, Dretske, 1986, and Neander, 1991). Groups may be able to represent the world through the representational states of their members, or may work in a different way. Bender and Koller (2020) have argued that contemporary Large Language Models don't actually represent the world, but just the language they have been trained on. But others have argued that there are various approaches that might solve the "grounding problem" for such systems (Mollo and Millière, 2023, Mandelkern and Linzen, 2024, and Lederman and Mahowald, 2024). In any case, there are obviously no well-developed theories for how the doxastic states aliens and angels might represent the world, and the creatures that do, do so differently enough that the question of how it is possible is outside the scope of this project.

But there are two main models of what representation might be like that have often been used—pictures and sentences. Their fit to the world seems to work in different ways. We usually think of a sentence as being either true or false, while a picture is usually more or less accurate in some ways and more or less inaccurate in others. Sentences can include information about non-visible parts of the world, and about things that are not the case, or that could or should be the case, that are hard to represent in a picture. But pictures can include information about precise visual properties and relationships that we have no words for.

When thinking about my own beliefs, it seems that they have aspects of each. But I think it's helpful to remember that these are just two different models, and that belief is likely different from either one. Furthermore, since I'm interested in doxastic states generally, including for creatures that might be quite unlike humans, it's worth keeping both models in mind (as well as any others) to overcome the presupposition that we know what a doxastic state is like just because we have one of our own.

Again, I will start by thinking about how we ordinarily *talk about* doxastic states. This ordinary language might be misleading or might miss some important aspects. I will eventually focus on a more formal account of doxastic states. But the account I focus on will still have many of the same features of ordinary reports, which might be a limitation.

1.3.2 Attitude reports

In ordinary language, when talking about someone's doxastic state, we report only a limited part of it. We don't try to describe their full mental model of the world, but just one aspect of it. We do this with a sentence like "Lois thinks Clark is at work", or "the bear doesn't realize that there's no food in the tent", or "my computer doesn't know which e-mail contains a virus", or "the FBI suspects the syndicate to be based in New York". A report of this sort involves three parts—the subject of the report ("Lois") is the being whose doxastic state is under description; there is a mental state verb ("thinks"); and there is a clause describing one part of the world ("Clark is at work").

In such a report, the verb and the following clause together specify some aspect of the doxastic state of the subject. The verb is said to specify the "attitude", and the clause is said to specify the "content" of that attitude. Some attitudes, like "know" and "realize", are said to be "factive", because a simple report with that verb doesn't just comment on the subject's doxastic state, but also introduces the presupposition that the content is true (even if the verb is negated—we wouldn't say the bear "doesn't realize" something unless we thought that thing was true). The grammar of the content clause can come in several different forms—it can be a full sentence (with or without the word "that" in front of it), or a noun plus an infinitive, or (if the attitude is factive) it can take the form of a question, in which case it seems to have the same meaning as the true answer to that question. (Only factive attitude verbs can take questions as their object, because factivity tells us which answer to the question is the one the subject must have the attitude to, while non-factive attitudes could be had to false answers as well the true one.)

The rest of this section will focus on understanding the content of attitudes, and will generally focus on the single attitude of "belief" when an attitude is relevant. However, in Section 1.4 I will consider what attitudes go into a doxastic state. In that section, I will give initial motivation for the Bayesian use of probability and a more precise description of the mathematics of probability. But the full argument won't occur until Chapters 2 and 3.

1.3.3 The nature of propositions

Gottlob Frege (1892) observed that although we report the content of the attitude with a sentence in some language, there's no reason to think that the sentence itself, or any sentence in any language, plays a role in the doxastic state. It may in the case of a person who speaks the language we are using to report the attitude, or in the case of a group or artificial intelligence that uses this language in its internal operations. But language and sentences are very unlikely to be relevant in the case of animals or aliens. And it also seems like we can perfectly well use English to report the attitudes even of a person who doesn't speak English, in which case the language we are using is unlikely to play any role in the doxastic state. Instead, philosophers use the term "proposition" for whatever the meaning of the content clause is. Even though there are several different grammatical patterns that can be used, and the attitude can be reported in any language, the proposition is whatever the shared content is.

One proposal about the nature of propositions is the view associated with Jerry Fodor (1975), involving the "language of thought". Various mental modules are in constant communication with each other and have some sort of representation akin to a language in which signals are passed back and forth. Fodor suggests that there is some mental structure we can crudely call the "belief box" and says that for a person to believe a claim just is for a sentence in the language of thought that expresses that claim to be in the belief box. This proposal takes the language of our attitude reports very literally. Although the clause in the attitude report may not itself be in the mind of the subject, some nearly equivalent sentence in the language of thought is.

This may or may not be accurate as a model of how the human mind works. It's very clearly correct about some artificial intelligences, but very clearly incorrect about others. One of the big advantages of modern neural nets over the "expert systems" of the 1980s is that the expert systems required explicit representation of huge amounts of knowledge for the system to work, while neural nets can extract and use information from training data by creating a distributed system of weights that allow it to effectively use this information without ever explicitly representing it (Dreyfus and Dreyfus, 1988). Whether the human mind (and those of animals or groups or aliens) works in something more like one way or more like the other or totally differently from either is primarily an empirical question, rather than a normative claim about what a doxastic state has to be like to do its job.

There is a somewhat more general view of propositions as structured entities, defended by King (1996) among others. On this view, whether or not a subject's mind itself contains "sentences in mentalese", there still is a meaningful sense in which attitudes involve relations to some sort of entity that has much of the structure of a sentence. This is parallel to some of Frege's contention that the mind somehow "grasps" structured "thoughts", which exist in a "third realm" separate from the mind or the physical world, but doesn't require as much contentious metaphysics. But as it turns out, arguments for and against structured propositions will be somewhat outside the scope of this book.

1.3.4 Propositions as sets

For our purposes, what is relevant is that, whether or not propositions have linguistic structure, much of the work they do only depends on what it takes for them to be true or false. This seems particularly true for the normative structure of doxastic states. The kind of action that a belief supports is the kind of action that is successful if the belief is true. Whether a belief is accurate or not depends only on whether it is true or false. Thus, for my purposes, it is usually most useful to identify propositions with their truth-conditions. (For an argument against this though, see Mahtani (2024).)

It is important to focus on truth-*conditions*, and not just truth. Truth by itself is sufficient for knowing whether a given attitude is accurate or not. Someone with a fortuitous true belief about which lottery ticket is the winner will be just as successful in this particular lottery as someone who has an extremely reliable and robust method for finding the winning ticket in various lotteries. Someone who correctly guesses that the rocket launch this afternoon will be successful is just as accurate as someone who has carefully studied the protocols and the past track record and determined that today's launch will succeed. But to understand the *difference* between the lucky person and the skilled person, and even to understand what it means that one set of beliefs is *about* the lottery and the other is about the rocket launch, we need to know not just whether the belief is *in fact* true, but under what conditions it would be true and under what conditions the subject would have it. The understanding of propositions that I will work with is that a proposition is to be identified with the set of circumstances in which it would be true.

Before I get into the nature of these "circumstances", I will introduce a bit of notation for sets. In most of the book, I will treat propositions as though they *are* sets of this sort, despite the reasons given in Section 1.3.3. I will usually use lower-case letters "p" and "q" to stand for propositions. If every member of p is also a member of q, we say that p is a "subset" of q, or "$p \subseteq q$", or that q is a "superset" of p, or "$q \supseteq p$". There is a specific set that I will denote as "$\emptyset$", that contains *nothing*, which represents any proposition that is true in no circumstances. For any proposition p, we have $\emptyset \subseteq p$ and $p \subseteq p$. For any two propositions p and q, there is a proposition $p \cap q$, the "intersection" of p and q, that contains exactly the elements that are in both p and q, and a proposition $p \cup q$, the "union" of p and q, that contains exactly the elements that are in either p or q. $p \cap q$ is the largest proposition that is a subset of both p and q, and $p \cup q$ is the smallest proposition that both are a subset of. If $p \cap q = \emptyset$, so there is nothing in both, then p and q are said to be "disjoint". All of this notation is standard for sets generally, whether it is sets of circumstances (which we identify with propositions), or sets of propositions, or sets of anything else. But it will be used most commonly for propositions in this book.

I will also use "$\neg p$" to stand for the negation of p and "$\bar{p}$" for the proposition that is true in all and only the possibilities in which p is false. Usually, $\neg p = \bar{p}$. But I will sometimes (as in Section 1.3.7) allow that a subject might be confused about possibilities in a way that causes problems for this. In this case, I will use $\bar{p}$ for the set-theoretic complement and $\neg p$ for the logical negation.

1.3.5 Possible worlds

The most common way philosophers think of circumstances, in these truth-conditions, is as "possible worlds". The idea here is that the actual world is just one of many possible worlds, and that what it is for something to be possible is for it to be true at one of the other possible worlds. We say that one set of propositions (the premises) logically entails another proposition (the conclusion) if the conclusion is true in every world that all the premises are true in, i.e., if the intersection of the premises is a subset of the conclusion.

Versions of this view go back at least to Leibniz, but the idea of a proposition as a set of possible worlds was developed further in the "modal" logics of necessity and possibility of C.I. Lewis and Langford (1932), Ruth Barcan (1946), and Saul Kripke (1963). While many of these authors were somewhat cagey about what exactly these "possible worlds" are, David Lewis (1986) was explicit that they are meant to be concrete universes just like the one we live in. On this picture, a proposition is just identified with a set of possible worlds, namely the worlds in which it is true. For one set of propositions to entail another is just for the intersection of the premises to be a subset of the conclusion.

This picture of a proposition as a set of worlds also allows a picture of the doxastic state as a unit, independent of how it is described by various attitude reports, at least in a context where belief is the only attitude under consideration. The proposal that is sometimes made is that the doxastic state can itself be identified with a set of worlds, perhaps the ones that the subject leaves open as possibilities, as illustrated in Figure 1.1. Then, there is no need for propositions themselves to be represented in the subject's mind—to attribute belief in a proposition to a subject is just to say that the subject's doxastic state is a subset of the set of worlds corresponding to that proposition. This account of a doxastic state (or a similar theory of the epistemic state built on knowledge rather than

A Possible Worlds Model of Belief

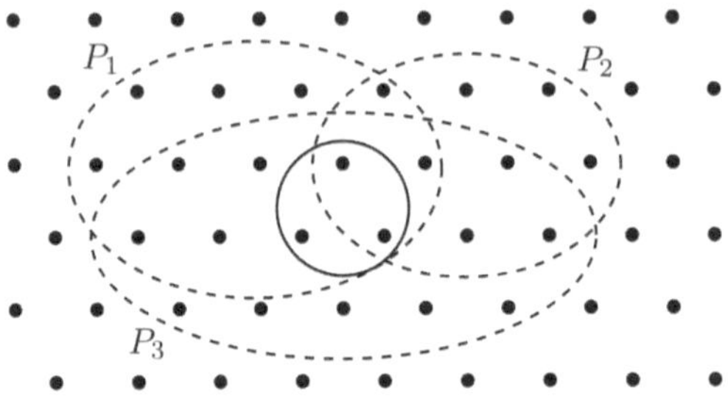

Figure 1.1 Each dot represents a possible world. The solid circle represents the subject's doxastic possibilities. The dashed ovals represent different propositions. P_1 and P_3 are believed, but P_2 is not.

belief) goes back at least to von Wright (1951) and Hintikka (1962), but was defended at length by Stalnaker (1984).

This model of belief has several interesting features. Because the believed propositions are the supersets of the doxastic state, if a subject believes a set of propositions, then they also believe the intersection of those propositions. And if a subject believes a proposition, then they also believe any superset of that proposition. Thus, if the subject believes a set of premises that logically entail a conclusion, then they automatically believe the conclusion. This is an appealing feature when we are talking about very simple logical entailments—it seems quite plausible that if someone believes the sky is blue and believes that grass is green, then they believe that grass and the sky are different colors. But this is less plausible when talking about more complicated entailments—Einstein already believed his theory of general relativity when he published it in 1915, but it wasn't until Schwarzschild proved that general relativity entailed that sufficiently large mass in a sufficiently small radius would result in a black hole that Einstein believed this.

Because a logically inconsistent set of propositions are not all true in any of the same worlds, this automatically means that a subject whose doxastic state corresponds to a non-empty set of worlds can never believe all of an inconsistent set of propositions. Or to put it contrapositively, a subject that believes each member of an inconsistent set of propositions must have a doxastic state corresponding to the empty set, $\emptyset$, and thus must in fact believe *every* proposition. Since most people do seem to have some contradictions among their beliefs, this would be a very bad consequence.

One major response that defenders of this proposal have made (developed originally by Lewis (1982), and more recently by Elga and Rayo (2021a,b)) is to say that a ordinary person's doxastic state is often "fragmented". Thus, although we might *say*, "David believes the train track runs north-south", and "David believes the street runs east-west", and "David believes the train track and the street are parallel", it is more precise to say, "one part of David believes the train track runs north-south" and "another part of David believes the street runs east-west" and "both parts of David believe the train track and the street are parallel". It's especially plausible that this is an accurate description of some AI systems and groups, but to me, this feels like it gets at something right about my own mind as well. I'll discuss this sort of fragmentation more in Chapter 8.

But there are also some deeper problems. Two sentences can denote different propositions only if it is possible for one of them to be true while the other is false. Thus, two necessarily true sentences automatically denote the same proposition, which is the set of *all* possible worlds. In particular, "$2 + 2 = 4$" and "$167 \cdot 263 = 70,221$" mean the same thing! Not only would this mean that anyone who believes one of these automatically believes the other, but in fact *everyone* who has a doxastic state would automatically believe *both*, since both are true in every world (and thus in every world that is part of the subject's doxastic state).

Furthermore, as Kripke (1972) argues, various *non*-mathematical claims may turn out to be necessary as well. Kripke argues that this is generally true for statements asserting the identity of two things that happen to be identical, or for statements expressing essential properties of something. Thus, "Mark Twain is Samuel Clemens" and "water is composed of hydrogen and oxygen atoms" also turn out to mean the same thing as "$2 + 2 = 4$" and are thus both automatically believed by everyone!

Thus, some sort of fix beyond possible worlds seems necessary to get at the relevant aspects of truth-conditions of propositions.

1.3.6 Two-dimensional semantics

One prominent approach to solve this, discussed in various forms by David Lewis, Frank Jackson, David Chalmers, and Robert Stalnaker, is by the method of "two-dimensional semantics" (Schroeter, 2017). Instead of identifying a proposition with a single set of worlds, they identify a proposition with a function from worlds to sets of worlds. The proposition still gives the meaning of a sentence, but the input world does something like determine which objects and properties the terms in a sentence are about, and the function then returns the set of worlds at which those objects have those properties.

Roughly speaking, the meaning of the sentence "Water is composed of hydrogen and oxygen atoms" would be a function that takes any world to the set of worlds at which *that* world's watery substance turns out to be composed of hydrogen and oxygen atoms. This sends the actual world to the set of all possible worlds (since Kripke has argued that the substance we think of as water must necessarily be composed of hydrogen and oxygen), but would take a different possible world (where the substance in lakes and streams has a very different chemical composition) to the set of *no* possible worlds. "$2 + 2 = 4$" corresponds to the function that takes any world to the set of worlds at which the numbers denoted by "2" and "4" in that world, and the operation denoted by "+" in that world, in fact bear the stated relationship. "$167 \cdot 263 = 70,221$" corresponds to a very different function. It will clearly be tough to work out the details of how this will work, but it's plausible that we will get different functions for these different sentences.

On this two-dimensional picture, the doxastic state can still be a set of worlds, but these worlds have to serve as both inputs and outputs of the function. That is, a proposition is believed by a subject if, whenever the function corresponding to the proposition is applied to a world in the doxastic state, the output is a set that is a superset of the doxastic state. A subject's doxastic state might be specific enough about the meaning of the terms in "$2 + 2 = 4$" that this sentence gets mapped to the set of all worlds at every world in their doxastic state. But the terms in "$167 \cdot 263 = 70,221$" might be hard enough to pin down (can you really picture the difference between 70,221 and 70,223?) that at some worlds in the doxastic state, the output is the set of all worlds, while at other worlds in the doxastic state, the output is the empty set. Necessity and logical relations are evaluated only by looking at the outputs of a proposition on the actual world, so the subject doesn't automatically believe the logical consequences of everything they believe or all logical necessities.

1.3.7 Impossible worlds

Another approach that I prefer is that of "impossible worlds". Versions of this approach have been developed by Berto and Jago (2018), Nolan (2021), and Soysal (2024) among others. On this approach, we don't take literally the idea of "possible worlds"—instead, we just have some abstract set of "worlds", both possible and impossible, and propositions are interpreted as sets of these entities, whatever they are.

Following Hacking (1967a), we can interpret (the relevant part of) a subject's doxastic state as a set of worlds, both possible and impossible, which I will call that subject's "doxastic possibilities". Even though necessities are true at all possible worlds, if a subject's doxastic possibilities include some impossible worlds, they might fail to believe those necessities. Even if some premises entail a conclusion, the subject can believe the premises without believing the conclusion, if their doxastic possibilities include some impossible worlds where the premises are true and the conclusion is false. This is illustrated in Figure 1.2.

The biggest problem with this framework is that, whatever understanding we have of possible worlds, *impossible* worlds are harder to understand. Even if we don't try to understand what they are, they are formally very unconstrained. There are decades of theory on what

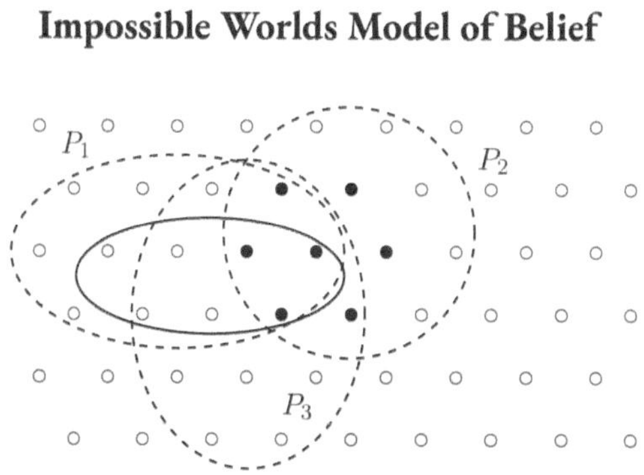

Figure 1.2 Black dots represent possible worlds and white dots represent impossible worlds. The subject's doxastic possibilities are inside the solid oval. P_1 entails P_3 because P_3 is true at every *possible* world that P_1 is. P_2 is a necessary truth, because it is true at all *possible* worlds. P_1 is the only one of these propositions the subject believes, because the others don't include all of the doxastic possibilities.

it takes for two sentences to have the same set of possible worlds as their meaning, but most impossible worlds frameworks give different accounts from each other. Does every proposition automatically include some impossible worlds as well as whatever possible worlds it contains?

Despite these difficulties, I favor this view for describing a doxastic state. What is relevant for a doxastic state, whether it is aimed at guiding action or aimed at accuracy, is whether it matches the world. No matter how it represents the world, there are various ways it treats the world as possibly being. These may not be possible in some objective sense—they may only be possible from the perspective of the subject with this doxastic state. To the extent that one can describe the doxastic state with attitude reports, all that matters about the propositions is which of these doxastic possibilities they are true in, and which they are false in. There may be better frameworks that combine some of the aspects of two-dimensional semantics with impossible worlds, like that of Kocurek (2021), but it is important that propositions be individuated in a way that is sensitive to their truth and falsity, but not just in possible worlds.

When talking about propositions as sets of these impossible worlds, I will follow a few conventions. The set-theoretic operations $\cap$, $\cup$, and $\bar{\cdot}$ will still work in the usual way—the proposition $p \cup q$ will be the set of possibilities in which either p or q is true, which might be different from the set of possibilities in which the subject takes "p or q" to be true, and similarly for the relation of $p \cap q$ and "p and q", and $\bar{p}$ and $\neg p$. As a reminder, "$\neg p$" will be the logical negation of p, which the subject might *take* to be true at some of the same possibilities in which p is, or might take not to be true in some possibilities where p is not, but "$\bar{p}$" is the *complement* of p, a proposition that the subject takes to be true at all and only the possibilities in which p is not. There may not be any expression in the subject's language that they take to work the way this $\bar{\cdot}$ operator works—but if every set of possibilities is a proposition, then this is a proposition.

1.4 WHAT ARE DEGREES OF BELIEF?

1.4.1 What are the attitudes?

So far, the main doxastic attitude I have discussed is that of belief. However, it's not essential to the role of a doxastic state that it be composed entirely of this one attitude. There could be many aspects and dimensions of the attitudes that go into a doxastic state. There might be different phenomenologies for different attitudes. Some attitudes could mix doxastic and bouletic elements, as might be the case with one person at a concert who *hopes that* the current song will be the last one of the show, and another person who *fears that* it will be.

However, as I will argue in Chapters 2 and 3, regardless of what sorts of attitudes do *in fact* go into the doxastic state of a given creature, only one dimension of those attitudes matters to the doxastic role. This single dimension is closely connected to the idea that contents only have one distinction, between the possibilities where they are true and the possibilities where they are false. In Chapter 2, I will discuss the pragmatic role of doxastic states in greater detail. I will argue that in determining whether there is a doxastically-relevant difference between the attitudes a subject has to two contents, all that matters is whether the attitudes lead the agent to prefer an act that depends for its success on the one content being true or an act that depends for its success on the other content being true. In Chapter 3, I will discuss the alethic role of doxastic states in greater detail. I will argue that in comparing attitudes to different contents, all that matters is which attitude is closer to treating the content as true and which is closer to treating the content as false. For a subject's doxastic state to play its role ideally, out of any two attitudes in this state, either one must be "higher" than the other or "lower" than the other or the two must be equal in this scale.

Thus, although the doxastic state might be composed of many types of attitudes, we can for doxastic purposes ignore all features of these attitudes apart from this ordering. If there are two distinct attitudes that are at the same position in this ordering, we can treat them as the same, since they are interchangeable for doxastic purposes. (Their differences may matter for the subject's non-doxastic rationality, but this is not our concern.) I will refer to the positions in this ordering, independent of any propositions the subject might have attitudes to, as "degrees of belief". If the attitude that a subject has to a proposition is a particular degree of belief, I will refer to that degree of belief as the subject's "credence" in the proposition.

1.4.2 How are degrees of belief related to each other?

Although "higher" and "lower" are the only comparisons that matter doxastically for a given *pair* of attitudes, there is more to the structure of degrees of belief when they are considered more than pairwise. It will turn out that there is a useful way to represent degrees of belief by real numbers on the scale from 0 to 1 in a way that helps clarify this structure. (In Chapter 5, I will consider the question of whether there might be some degrees of belief that differ "infinitesimally" from each other, so that a somewhat finer scale than the real numbers might be useful. But the overall structure of this scale will still be usefully represented as running from 0 to 1.)

I will not go into the details of how this representation gets off the ground until Chapters 2 and 3, where the versions for pragmatic and alethic roles are given. Those Chapters will also rehearse standard arguments showing that a subject's credences in various propositions do their jobs best only if they satisfy the axioms of probability theory:

- Non-negativity: A subject's credence in any proposition is non-negative.
- Normalization: If a proposition is true in all of a subject's doxastic possibilities, then their credence in that proposition is 1.
- Additivity: If there are no doxastic possibilities where two propositions are both true, then the subject's credence in a proposition that is true in all and only the possibilities where either of them is true is the sum of their credences in those two propositions.

(In Section 5.3 I will briefly discuss extensions of Additivity to infinite sets of propositions, but this is not relevant for most of the concerns of this book.) I will often use capital letters (usually "P") to stand for a particular probability function (which may represent a subject's credences) and lower-case letters (usually p and q) to stand for propositions. With this notation, the axioms become:

- Non-negativity: $P(p) \geq 0$
- Normalization: $P(p) = 1$ if p includes all doxastic possibilities
- Additivity: If p and q are disjoint, then $P(p \cup q) = P(p) + P(q)$.

These axioms imply all the other mathematical features of probability, including things like the fact that if $p \subseteq q$, then $P(p) \leq P(q)$. If there are only finitely many possibilities, then each one has a probability, and the probability of any proposition is just the sum of the probabilities of the possibilities in which it is true. Many mathematical textbooks of probability are filled with examples involving coin flips and dice rolls and shuffled decks of cards, and an assumption that each possibility has the same probability, such that probabilities can be calculated by counting the fraction of possibilities in which a proposition is true.

In Chapter 6, I will also introduce the idea of "the conditional probability of p given q", notated as "$P(p|q)$". This is mathematically defined as $P(p \cap q)/P(q)$, when $P(q) \neq 0$. It is something like the fraction of q possibilities in which p is also true. (How to interpret conditional probability when $P(q) = 0$ is beyond the scope of this book, but it is something I have written extensively about elsewhere, particularly in Easwaran (2008) and Easwaran (2019).) This quantity plays important roles in determining how subjects should best update their credences as they learn new things. In particular, if a subject has credences given by a probability function P and learns some proposition q and nothing else, the rule of conditionalization says that their new credence in any proposition p should be equal to $P(p|q)$. Standard justifications of this rule are presented in Chapter 6, as well as some considerations about the kinds of cases in which alternative revision procedures might be more reasonable.

"Bayesianism" is often taken as the doctrine that degrees of belief satisfy the probability axioms, and that they update by something like conditionalization. The doctrine that degrees of belief satisfy the probability axioms is often referred to as "probabilism", and in Chapters 2 and 3, I will primarily discuss the arguments for probabilism on the two functional accounts of the doxastic state. Only in Chapter 6, do I consider the full arguments for Bayesianism, though in Chapter 7 I note the ways that these parts might be separable.

The word "probability" is often associated with various concepts of chance or randomness. In textbook examples involving gambling, we often think of there being some indeterminacy in the world that the probability is measuring. However, when we are talking about degree of belief, the propositions can be about anything, including perfectly deterministic matters. What matters for probability in the Bayesian sense of credence is just the subject's certainty or uncertainty, not whether there is some objective chance. (Though subjects who believe that there is objective chance about something will often reflect this in their credences.)

Throughout all this, it is helpful to occasionally recall that the real numbers are not themselves the mental states. The real numbers are convenient representations for degrees of belief, which are equivalence classes of doxastic attitudes under the doxastically important relations. Getting clearer on all this will be the focus of Chapter 4.

1.5 HOW DOES "DEGREE OF BELIEF" RELATE TO "BELIEF"?

The doxastic attitudes that are relevant for Bayesianism are degrees of belief. But as mentioned earlier, belief also appears to be a doxastic attitude of a similar sort. Many philosophers in recent years have addressed the question of how these two attitudes relate. It could be that one of these attitudes is just an illusion, and that actually the other fully constitutes the doxastic state. Another subtly different possibility is that although both of these attitudes might be correct descriptions of part of a subject's doxastic state, one of them might just be a way of talking about patterns in the other, so that the other is more fundamental. (Compare: evolution proceeds both by Darwinian natural selection and Mendelian genetics, but the Mendelian genes are the constituents of Darwinian inherited characteristics—in the early 20th century these were seen as competing theories until one was reduced to the other.) Another possibility is that the attitudes of belief and of degree of belief are both reducible to the same third thing. Or finally, it could turn out that belief and degree of belief just are two separate aspects of the doxastic state. In the latter case, there may still be normative connections between the attitudes, and there has been debate about whether one of the attitudes drives the normative features for the other or whether the normative connections apply to the two symmetrically.

Because of the great diversity of possible subjects for doxastic states—not just humans, but the "five A's" as well—I don't think there is any prospect for a single correct answer about the relationship between the attitudes of belief and degree of belief. It's very likely that different creatures could have different relations among the two, just as I think that some creatures could have doxastic states that play a pragmatic role and others an alethic role. So the philosophers engaged in this debate must primarily be addressing the empirical question of what the relationship is between these attitudes for humans. But I will summarize some of this debate here because it might help understand the ways that degree of belief could be constituted in different types of creatures, and get clearer on the roles that degree of belief plays. (For a more detailed summary of much of this literature, refer to Jackson (2020).)

1.5.1 Degree of belief is primary

Among philosophers who have explicitly considered the question of the relation of degree of belief and belief, the most popular view has probably been that degree of belief is the more fundamental notion.

1.5.1.1 Eliminativism about belief

Bruno de Finetti (1931) suggests that the goal of a doxastic state is just to make predictions about the world. Due to the skeptical considerations of Descartes and Hume, he thinks that these predictions must always come with varying degrees of uncertainty. Thus, even if we do have an attitude of belief, it is unnecessary, and more idealized reasoners would just work with greater or lesser degrees of belief.

Frank Ramsey (1926) comes to a similar view for behaviorist reasons. He thinks that the way to understand doxastic states is through behavior, and he thinks that the only doxastic states that can be understood through behavior are degrees of belief. Thus, degree of belief is the only real doxastic state. Richard Jeffrey (1970) says that Ramsey's analysis of degree of belief "sucked the marrow out of the ordinary notion [of belief], and used it to nourish a more adequate view" (p. 172). Even if belief does exist beyond degree of belief, it is degree of belief that explains all the interesting doxastic phenomena.

1.5.1.2 The Lockean thesis

Some more recent philosophers have taken less radical versions of this view about the priority of degree of belief over belief. Richard Foley (1993) claims that degree of belief and belief both exist, but that belief *just is* sufficiently high degree of belief. That is, rather than belief being a separate state from degree of belief, the idea is that there is some threshold t, such that believing p just is having credence greater than t in p. (By a simplistic reading of a certain passage from John Locke (1690), this view has come to be known as the "Lockean thesis".)

An important and interesting consequence of this view of belief is that it avoids the strong closure conditions implied by the version on which belief involves truth in all doxastic possibilities. Consider the Preface Paradox (Makinson, 1965). A scholar has earnestly researched and written up a history book. It seems correct to say they believe each claim that they make in the book. However, in the preface, after thanking various colleagues, friends, and family for assistance, they remark, "the errors that remain are my fault alone, and not anyone else's". It seems correct to say they believe that the book contains some errors. On a view in which belief requires truth in all doxastic possibilities, this would mean that every doxastic possibility for this person involves the truth of every claim in the book, but also involves the book containing a mistake. This might be possible, depending on what these doxastic possibilities are like. But Foley is able to give a more satisfying account—none of these propositions is true in *every* doxastic possibility, but all are true in the overwhelming majority of the doxastic possibilities. It's natural for a person to have high degree of belief in something despite allowing some possibility in which it is false. If the book contains 1000 claims, then each claim could be true in 1000 of the doxastic possibilities and false in 1, while the preface claim (that the book contains an error) is also true in 1000 possibilities and false in 1.

It's not quite as clear whether Foley's view deals well with the formally similar Lottery Paradox (Kyburg, 1961). In a lottery with 1000 tickets, one can have a high degree of belief (specifically, .999), of each ticket, that it will lose, while yet having high credence that there will be a winner. The natural model of doxastic possibilities, in which there is one doxastic possibility for each ticket to win, and some ticket wins in every possibility, yields judgments that Kyburg wanted for the lottery paradox, that existing models of belief at the time could not account for.

However, contemporary epistemologists have raised some worries for it (Ryan, 1996 and Nelkin, 2000). Some have said that although the Lottery Paradox is formally similar to the Preface Paradox, an appropriate theory of rational belief should treat them differently. In the case of the lottery, the fact that there is a psychologically salient doxastic possibility in which the ticket fails to lose seems to make many people say they don't believe the ticket will lose (even if the lottery is extremely large, such that this degree of belief is extremely high).

When tidying up the house, if one finds an old lottery ticket and believes that the ticket is not eligible to win (perhaps because the lottery is dated several years in the past), one is willing to discard the ticket. But when finding a ticket for a lottery that is about to be conducted, again, no matter how large the lottery is and how high the degree of belief that the ticket won't win, if the prize is sufficiently large, it looks like one shouldn't just discard the ticket until the lottery is conducted and the winner is known. If belief is thought to be sufficient for action, then this suggests that no threshold on degree of belief is high enough to suffice for belief. However, Locke (2014) argues that a variable-threshold version of the Lockean thesis might hold, where the threshold needed for degree of belief to count as belief depends on the practical stakes of available actions.

In Section 1.3.5, we showed that if belief is defined by truth in some particular non-empty set of possible worlds, then belief is automatically closed under logical consequence, and a subjects set of beliefs is automatically logically consistent. Some philosophers have been interested in developing a reduction of belief to degree of belief that has these features, that doesn't require a believed proposition to be true in every doxastic possibility. Douven and Williamson (2006) argue that any "purely formal" reduction of this sort will in some cases be trivial. In particular, they show that for a large enough set of possibilities, if a subject has equal degree of belief in each possibility, then there must be no proposition that the subject believes other than the propositions true in every doxastic possibility.

1.5.1.3 Other reductions

However, several recent authors accept that their reduction of belief to degree of belief will be trivial in the case where all possibilities have equal degree of belief and are interested in giving non-trivial reductions of belief to degree of belief just in cases where different possibilities have different degree of belief. One version of this idea was developed independently by Arló-Costa and Pedersen (2012) and Leitgeb (2014) (though Leitgeb's version has received more uptake in the literature). On this picture, there is a "belief core" such that a proposition is believed if and only if it is true in every possibility in the core—but the core can be identified with purely probabilistic considerations, such that every possibility in the core has higher probability than *all* possibilities outside the core considered together. Not only does the concept of belief here non-trivially satisfy closure and consistency, but it also satisfies a "stability" condition. Once a subject believes a proposition, if they update their credences by Bayesian conditionalization on propositions that don't logically contradict anything they believe, they preserve all their previous beliefs and add more. The resulting analysis of belief satisfies the full belief revision axioms of Alchourròn et al. (1985).

Another reduction has been provided by Lin and Kelly (2012). They motivate their reduction by looking for a reduction of belief to degree of belief, together with a natural belief revision rule that operates just at the level of belief, and produces the same updates as Bayesian conditionalization on the underlying degrees of belief. However, they argue that full stability is implausible. In some cases, one might give up a belief p, not because one learned anything logically inconsistent with any of one's beliefs, but just because one rules out a particularly plausible way for p to obtain. I have two friends I'm planning to go out with for lunch today, and I start out believing that one of them has brought their car to campus, without specifically believing of either one of them that *she* brought her car today. When I learn that the one who more often brings her car didn't bring it today, I give up the belief that one of them has brought a car, even though I didn't learn anything I had specifically disbelieved. Lin and Kelly show that their axiomatic approach gives a reduction of belief to degree of belief that depends on the *ratios* of degrees of belief in particular doxastic possibilities rather than the absolute degree of belief. Their approach yields a non-trivial set of beliefs in an even wider range of possible credences than Arló-Costa and Pedersen (2012) or Leitgeb (2014).

1.5.2 Belief is primary

Although philosophers who explicitly consider the relationship between belief and degree of belief have usually leaned toward the view that degree of belief is more fundamental, there are some who have taken the opposite view.

1.5.2.1 Eliminativism about degrees of belief

One particularly forceful version of such an argument has been given by Gilbert Harman (1986). Harman actually argues that degree of belief is outside the computational capacity of creatures like us. As he points out, belief allows one to discard all the possibilities in which a proposition false. Having a greater number of beliefs can in some sense simplify one's cognitive processing. But with degree of belief, one often doesn't rule anything out. Having degrees of belief in more propositions makes the complexity of the doxastic state grow, rather than shrink, as one needs degrees of belief for all the interactions of these propositions as well. Harman specifically argues that the number of possibilities and number of degrees of belief grows exponentially with the number of propositions considered. Adding a new proposition doesn't just add two new objects of credence—it *doubles* the total number of possibilities, because all existing possibilities must be considered both in a form where the new proposition is true and in one where it is false.

Of course, the success of this argument depends on empirical facts about the processing power of the human brain. Even though humans are bad at explicitly tracking lots of mathematical facts, some features of our mind do a good job of some of it. We are bad at explicitly solving differential equations in three-dimensional space, but we are good at catching balls and frisbees whose motion is tracked by these equations. In fact, many of our beliefs about how we move our eyes, hands, and bodies while trying to catch a ball are robustly false (Reed et al. (2009)). The methods we use to do these things are not perfect, but they are good enough for practical purposes, and the same could easily be true of our management of degrees of belief. I will address this argument at greater length in Section 4.2. In any case, this objection clearly doesn't apply to all possible subjects of doxastic states—even if individual humans don't have the capacity to manipulate full sets of credences, it doesn't necessarily pose problems for artificial intelligences, formalized groups like the statistical community at large, or alien or angelic intelligences.

1.5.2.2 Degree of belief as belief about probability

But even without the Harman view that humans *don't* have degrees of belief, many philosophers have argued that humans have degrees of belief by means of having particular beliefs.

The most popular way to do this has been to argue that facts about degrees of belief are constituted by beliefs in probabilistic contents. Consider the fact that my degree of belief that it will rain here some time in the next week (p), while low, is not as low as my degree of belief that there is currently living life on the surface of Mars (q). I might express this relation by saying something like, "I don't think it's very likely that it will rain this week, but I think it's more likely that it will than that there's life on Mars". Taking the surface grammar of this sentence at face value, and assuming that the verb "think" expresses the attitude of belief, it looks like I am not fundamentally saying anything about degrees of belief in p and q, but am rather directly attributing belief in two complex propositions: "it's not very likely that p" and "it's more likely that p than that q". Many philosophers have argued that this is the best way to interpret degree of belief—it is not really a distinctive attitude, but is instead just ordinary belief in a proposition that involves some concept of probability. For instance, versions of this view are defended by Moon and Jackson (2020) (note that Jackson herself does not endorse this view, see Jackson (2019, 2022), but Moon does;

Jackson merely signs on to defend this view against certain objections that she thinks are flawed) and Buchanan and Dogramaci (2024).

The biggest challenges for such a view involve identifying the concept of probability involved in the content of the belief, and arguing that all humans that have degrees of belief can grasp the relevant content. While it's plausible that I might believe that the weather is indeterministic, such that there is a non-zero, but low objective chance that it rains this week, I definitely don't believe this about life on Mars. I believe that there either is or is not life on Mars, and there is no chance about this, so my belief that it's not very likely that there is life on Mars can't be cashed out as a belief about objective chances. Instead, most philosophers who defend this reduction of degree of belief to belief argue that the relevant concept of probability is some sort of objective "evidential probability". On this account, for it to be not very likely that there is life on Mars is for there to be some sort of relationship between my "evidence" and the proposition that there is life on Mars that can be cashed out in probabilistic terms, that gives that proposition a low probability. This "evidential probability" relation can't be interpreted as providing the degree of belief that an ideally rational subject would have in the proposition, given the evidence, because this relation is meant to explain the concept of degree of belief, rather than being explained in terms of it. Buchanan and Dogramaci (2024) say to "read this paper as not just blindly assuming the hypothesis that there are evidential probabilities, but as showcasing the elegant theory of credence it can give us."

Sarah Moss (2017) has given a formal semantics of belief ascriptions that seems to go farther than this. She suggests that *all* contents of belief are actually best interpreted as being not sets of worlds, but rather sets of probability functions. The content of my belief that it's more likely to rain this week than that there is life on Mars is the set of probability functions on which the former proposition has higher probability than the latter. But the content of my belief that Paris is the capital of France is also a set of probability functions, namely the ones on which it is highly probably that Paris is the capital of France. However, in understanding the correctness conditions of these reports, Moss makes it clear that the probability function involved is not any objective chance or evidential probability function. Instead, the most natural interpretation of Moss's thesis is that the "attitude" of belief that she postulates is just a tool of formal semantics. Each thinker just has a doxastic state that consists of a probability function, and a belief report is true of such a thinker if and only if their probability function is contained within the set that

Moss says is the denotation of the sentence they are said to believe. This ends up being a view on which belief reduces to credence, rather than credence reducing to belief.

In fact, Yalcin (2012) has argued that the surface grammar of belief reports about probability—or even bare assertions like "it's more likely that it will rain this week than that there is life on Mars"—is actually misleading. He develops an "expressivist" semantics for these sentences on which they just express facts about the speaker's degrees of belief. If correct, this undercuts the entire idea of reducing degree of belief to belief by means of probabilistic contents.

1.5.2.3 Degree of belief as pattern in beliefs

I have developed one other means by which a creature might have degrees of belief by means of having beliefs in my Easwaran (2015). Rather than grounding each particular degree of belief in a particular belief about a probabilistic content, I sketch a way that degree of belief talk could be grounded in the beliefs of a creature, without any particular degree of belief ascription resting on a particular belief. This particular approach is based specifically on the alethic role of doxastic states, and I will discuss it in greater detail in Section 3.2.

But the basic idea is to reverse the Lockean thesis discussed in Section 1.5.1.2. Rather than saying that belief just is credence above some threshold, we say that credence just is any probability function that puts all the beliefs above the threshold and all the propositions that aren't believed below that threshold. By considering the role of belief in aiming at the truth, we argue that there always will be some probability function that satisfies these constraints, and by looking at the variety of propositions that a subject has beliefs about, we argue that any two such probability functions will be very similar to each other. The probability function doesn't need to correspond to anything real, the way that "evidential probability" needs to in order to make the contents of belief meaningful. Instead, the probability function is just a mathematical tool for representing the beliefs of this particular subject.

1.5.3 Belief and degree of belief both exist

While many philosophers have tried to either eliminate one of belief and degree of belief or reduce one to the other, a more popular position recently has been to say that both attitudes are part of the doxastic state of actual humans, and that there is some amount of independence between the two. But in order to justify this sort of view, one needs to explain what role each of the two attitudes plays. In this section, I will describe several proposed roles for belief. However, in the rest of the book, I will primarily focus on degree of belief, because that is the only class of attitudes directly relevant for Bayesianism.

One possibility is that belief plays a similar sort of alethic or pragmatic role to degree of belief. On the ways of fleshing out the alethic role that have been investigated, belief and degree of belief end up naturally collapsing into each other. As I showed in Easwaran (2015), if a subject has beliefs that do a good job of achieving their alethic role, there is also a useful description of the subject on which they have degrees of belief that play the standard alethic role. And Dorst (2019) shows that if the subject also has degrees of belief that play the pragmatic role while having beliefs that aim at the truth, then there will be some threshold t such that the subject's beliefs ought to relate to their degrees of belief by the Lockean thesis.

When the role of belief is pragmatic, more complexities have been discovered. One natural thought is that believing that p play the role of allowing one to act as though p. This is the idea that makes it seem as though one shouldn't believe that a lottery ticket is a loser solely on statistical grounds—acting as though the ticket is a loser means throwing it out when tidying up, but if the prize is large enough, it's worth hanging onto the ticket at least until the winner is called. However, it may be reasonable to believe each claim in a book, as long as there aren't any high-stakes actions one is considering, even though one also believes that there is at least one false claim in the book. The phenomenon in which changed stakes can make belief (and knowledge) disappear is known as "pragmatic encroachment", and was first discussed by Fantl and McGrath (2002), though the first discussion of how it applies specifically to belief came from Weatherson (2005). A precise statement of the impact of this feature was give by Locke (2014).

This ends up tying belief quite closely to degree of belief, but as mentioned in Section 1.5.1.2, there is no particular degree of belief, such that belief in a proposition corresponds to having that degree of belief in it—instead, the threshold varies based on what kinds of actions are available. My degree of belief that my house won't burn down this year is well over .999, and so normally I believe this proposition. (Even in the awful fires of January 2025, fewer than 1/1000 of the homes in California were burned, and my home is not in the wildland interface where most destructive fires occur.) But when I was choosing between available insurance plans and I saw that a change in the fire coverage only affects the price of the overall plan by a few dozen dollars per year, this belief disappeared, and I made the choice on the basis of my degree of belief. If beliefs govern my choice of actions, then when new actions become available, my beliefs might change. But the fact that my degrees of belief remain roughly constant throughout all this, and successfully govern my actions at all times, might lead one to prefer to focus just on degree of belief for practical action, and not belief. (Lin (2013) has attempted to give a different analysis of the pragmatic significance of belief, though it still ends up relating it very closely to degree of belief.)

A related role that has been proposed as a more distinctive one for belief, is to ground moral attitudes. Buchak (2014) considers a case in which someone has been wronged, but doesn't know for sure by whom. She claims that if the person believes that a particular person is responsible, then it would be appropriate to resent that person and attempt to hold them legally liable, but it wouldn't be appropriate to do so merely on the basis of high degree of belief. For instance, imagine a charter bus has damaged the person's parked car, and they don't know which bus company is responsible. Buchak notes that it seems inappropriate to resent the Blue Bus Company or try to hold them liable merely on the basis that they operate 90% of the charter buses in town while the Green Bus Company operates the other 10%. However, she notes that it does seem appropriate to resent the company and try to hold it liable, if an eyewitness reports seeing a blue bus scrape the car. Importantly, many people feel this is appropriate even if the eyewitness is less than 90% reliable at identifying blue and green buses in the lighting conditions at the time.

Whether or not one agrees with Buchak's account of that case, philosophers have presented other cases that motivate similar "moral encroachment". Rima Basu (2018) argues that even without grounding other moral attitudes like resentment and liability, there is moral significance to belief itself, separate from degree of belief. She argues that believing of a person that they must be a waiter rather than a guest or that they must be a bad tipper merely on the basis of their visible race or gender identity is morally wrong. She further argues that this moral wrongness in fact makes the belief epistemically unjustified. Some significant further evaluations of this thesis have been made by Moss (2018), Gardiner (2018), and Jorgensen Bolinger (2020).

Another role for belief that has been emphasized at various points is that belief seems to play a role in providing an overall consistent picture of the world. This is importantly different from what degree of belief does—after all, in the case of the lottery, one has a high degree of belief, of each ticket, that it will lose, but also has a high degree of belief that one of the tickets is a winner. Despite all of these propositions having high degree of belief, they can't all be true together. This is part of what motivates Ryan (1996) in her approach to the lottery, saying that no degree of belief is high enough to be sufficient to ensure belief.

Interestingly, in recent years, some philosophers have argued that even though no degree of belief is high enough to ensure belief, nevertheless, belief is compatible with quite low degrees of belief. Hawthorne et al. (2016) argue that in fact there are some things we believe that we have relatively low degree of belief in. Dorst and Mandelkern (2021) and Holguín (2022) develop a formal theory of how this derives from the idea that belief gives our best consistent picture of the world. On this view, what one believes depends on what questions one is considering. To believe something is to treat it as the best answer to that relevant question. How good a proposition is as the answer to a question depends both on how high your credence is in the proposition, and also on how important the question is to you, including both how important it is to have the correct answer, and how important it is to avoid any incorrect answer. Depending on these factors, the best answer to the question might be the precise answer one has highest credence in ("horse C will win the race"), or it might be a disjunction of several answers, all of which one has higher credence in than any others ("either horse C or E or G will win the race"). If there are many possible answers, some of which are much better than the alternatives, it could still be that one believes something that one has relatively low credence in. This is reminiscent of an argument James (1896) gives for why it can be reasonable for a modern person to believe in a particular religion, even though there are many alternatives, and not much evidence to favor one over another—as long as the possibility of believing the truth is important enough, while they aren't so concerned with avoiding false belief.

One last role I consider for belief is the role of simplifying reasoning for limited beings. While Weatherson (2005) and Locke (2014) suggest that belief has a pragmatic role that relates it to degree of belief above a threshold, Ross and Schroeder (2011) take a slightly different interpretation of these cases. They argue that the role of belief is not directly in guiding action—instead, they argue that the role of belief is to exclude consideration of certain possibilities when reasoning. They consider a case where Sarah has degree of belief .95 that the peanut butter sandwich is on the left and the almond butter sandwich is on the right, and her nephew asks for the almond butter sandwich. In the case where the nephew is allergic to peanuts (but not almonds), they follow Weatherson and Locke in saying that Sarah should stop and check again before giving her nephew a sandwich, so she shouldn't act as though she believes that the sandwich on the right has almond butter. But they note that even when Sarah knows her nephew isn't allergic, and just has a mild preference, there are two ways that she might act. They say that for her to *believe* that the sandwich on the right has almond butter is for her to just give this sandwich to her nephew without a further thought. But they note that she might nevertheless *consider* the possibility that she is wrong, before deciding that this is unlikely enough, and the stakes low enough, that she just gives the sandwich without checking. They claim that "belief" marks this distinction between considering alternate possibilities and not considering them, regardless of whether or not it makes a difference to action.

Drawing this distinction between "considering" and "not considering" is different from most of the other roles considered so far, because it is one of the few that depends on the details of how the mind actually functions. (The role in grounding moral attitudes may as well.) Most of these other roles just depend on having certain dispositions to act, or to answer questions, or the like. Those other roles seem just as applicable for subjects such as corporations or computers or animals. But this one may need extra work to translate it to other kinds of creatures.

A related idea has been developed by Justin Dallmann (2015, 2017). He takes Harman's objection seriously, as well as taking the internal workings of the mind as significant in the way Ross and Schroeder do. While Ross and Schroeder (2011) just identify the elimination of consideration with belief, Dallmann gives an argument for why creatures of a certain sort would want an attitude that did this work. He considers a model in which a subject is managing a doxastic state that deals with many different propositions and many different types of evidence for them, with a limited capacity for processing that evidence in an expeditious manner. Dallmann notes that such a creature will sometimes fail to consider the import of certain pieces of evidence, just by virtue of being overwhelmed. He argues that, rather than allowing this to happen randomly, it would be better for the creature to actively ignore evidence relevant only to propositions that are already extremely well-supported, particularly if the evidence that comes in for this proposition is particularly frequent (so that ignoring it frees up a lot of processing power) and usually continues to support the proposition even more strongly. Taking such propositions out of the heavy processing demands of degree of belief, by moving them into another category of belief, helps the creature manage its attitudes to other propositions more effectively.

This short survey of views on the relation between belief and degree of belief is inherently limited. But it is helpful to have some such consideration to get a clearer sense of what precisely is included in "degree of belief", which is the core attitude that Bayesianism is about.

1.6 CONCLUSION

Bayesianism is a view in epistemology, about doxastic states. I think it is useful to focus on an account of doxastic states that focuses on what function they play for the subject that has them, so that we can make normative claims that are justified by the subject's own functions. I consider both the pragmatic function of governing effective action and the alethic function of aiming at the truth. In either case, the doxastic state is the representational feature of the creature that represents the world with a mind-to-world direction of fit. Humans are the paradigm doxastic subjects, but my aim in this book is to show why Bayesianism is useful for any such doxastic subject, including animals, aliens, artificial intelligences, associations of people, and angelic intelligences, whether or not any particular such subject perfectly lives up to the Bayesian norms. I make the substantive assumption that the doxastic state can be described by attitudes of the subject to individual contents, rather than in a purely holistic way. But as I will argue in later chapters, the contents can usefully be understood just in terms of what it takes for them to be true or false, and the relevant attitude of degree of belief can be understood on a single linear scale, aiming toward truth at one end and falsity at the other. "What it takes" for them to be true or false is just understood in terms of what set of "doxastic possibilities" they are true in. And as I will argue in later chapters, the Bayesian ideal allows for these degrees of belief to be represented with real numbers from 0 to 1, that satisfy the probability axioms, and that update in ways illuminated by conditional probability.

Many philosophers are more interested in a distinct doxastic attitude of "belief", that may be connected to degree of belief in one of a number of ways. Many philosophers are more interested in non-doxastic aspects of the functioning of the states that play these roles, whether for humans or for other creatures. But Bayesianism is a theory of the doxastic role of degrees of belief, regardless of what kind of creature has it.

In Chapters 2 and 3, I will discuss the pragmatic and alethic roles for degree of belief and show how they both lead to the probability axioms, as well as discussing various issues that have arisen for each. In Chapters 4 and 5, I will discuss the use of numbers to represent degrees of belief, to alleviate worries that numbers are a misrepresentation of the mind. In Chapter 6, I will discuss the way in which a doxastic state should update over time to complete the justification of Bayesianism. And in Chapters 7 and 8, I will return to several broader normative issues.

RESEARCH QUESTIONS

1. Are there other roles a doxastic state could play beyond the pragmatic or alethic, that would make it count as a doxastic state?
2. Is there a useful way to conceive of an overall doxastic state that doesn't factor into attitudes toward individual propositions?
3. Are there important features of the contents of doxastic attitudes that aren't captured by their truth conditions in doxastic possibilities?
4. Is there a better model for doxastic possibilities than impossible worlds or two-dimensional semantics?
5. How essential are non-doxastic aspects of various doxastic attitudes for understanding the doxastic states of actual humans?
6. For actual humans, what is the relationship between full belief and degree of belief—does only one really exist? do they both exist with one more fundamental? or are they in some substantive sense co-equal?
7. Similarly for other types of actually existing doxastic subjects—animals, groups, artificial intelligences, etc.
8. Are there other ways a subject could have doxastic states, other than as beliefs or degrees of belief?

CHAPTER 2

Belief guides action

THIS CHAPTER IS an introduction to the pragmatic role for degree of belief as a guide to action. The central idea of belief and desire working together to guide action goes back at least to David Hume in the 1780s, but it only began to be worked out formally in the work of Frank Ramsey and Bruno de Finetti in the 1920s (though some of these ideas are already suggested by Butler (1736) and Bayes (1763)). In the century since Ramsey and de Finetti, it has been treated as the main philosophical foundation for Bayesianism. However, in the past few decades, there has also been significant interest in alethic alternatives, which will be discussed in Chapter 3.

To spell out the pragmatic role for degree of belief, we have to say more about action. What it is for a behavior to count as *action* in the relevant sense, rather than mere reflex or other sort of behavior, is for it to be appropriately caused by representational states. There is of course a lot more detail on this discussion in the literature on philosophy of action, including lots of detail about what "appropriately caused" might mean (Piñeros Glasscock and Tenenbaum, 2023), but for my purposes, this crude account will be sufficient. I am not particularly interested in what is most distinctive of human action in the most precise sense, but rather in what unifies it with many other related sorts of behavior of many other creatures. The idea is that an agent—a creature that acts, from the Latin "agere", meaning "to act"—has some representations of the world, and it engages in some behaviors. It has some representation of how the world should be, and some representation of how the world is, and uses them to produce a behavior that (hopefully) brings the world from the latter toward the former. On this picture, an agent's doxastic state and their bouletic state are two representational features

DOI: 10.4324/9781315626918-2

of the agent that work together to govern some significant parts of their behavior, with the two parts having distinctively different roles.

Belief, desire, and action are defined by their relationships to each other. Belief and desire are distinguished from each other, as well as from other representational states of the agent, by the contributions they make to action. Action is distinguished from other behavior by the way that it is governed by belief and desire. My walking to the Chinese restaurant is caused by my representations of myself being full in the future, and of what food the restaurant has served in the past, and of the restaurant being open now and still serving similar dishes. Thus, we think of this walking as an action, the representation of myself being full in the future as a desire, and my representations of the past food at the restaurant, and its current openness, as beliefs. Other representations, like imaginations and suppositions, don't affect behavior in the same way, which is what distinguishes them from beliefs and desires. Other behaviors, like the kicking reflex when the tendon near my knee is hit aren't controlled by representations, aren't controlled by representations in the same way, which is what distinguishes them from actions.

However, this mutual definition sometimes leads to an ambiguity. After I eat, my stomach undergoes some complex motions, churning its contents and secreting various fluids, processing the food into simpler nutrients that can eventually enter my bloodstream. This process is controlled by various hormones like ghrelin and leptin that signal my body's need for energy and by specialized parts of my nervous system that are localized to the digestive tract, which sense the presence of food and move the gastric muscles appropriately. We don't think of the behaviors of my digestive system as "actions", because they aren't intentionally produced by my beliefs and desires. However, there might be an alternative interpretation. We could consider the representational states of the enteric nervous system to be "beliefs" that food is present. We could consider the hormonal system involving ghrelin and leptin (which is the main target of modern weight-loss drugs like semaglutide and tirzepatide in medications like Ozempic and Mounjaro) to represent "desire" for more energy. On this interpretation, digestion could well count as an "action", since it is behavior that is produced and modulated by "beliefs" and "desires". We don't normally think this way, because we normally identify a person with the beliefs and desires that are more located in the brain, rather than the ones located in the hormonal systems and the enteric nervous system. But if we are willing to think of animals, artificial intelligences, and groups as agents, then we might think of my digestive system as another agent located within my body.

This chapter outlines how considering the way degrees of belief and desire interact to produce action leads naturally to the probabilism degrees of belief. Not every "agent" will do a great job of living up to these norms (and the associated norms for update of degree of belief described in Chapter 6), but for the ones that do, Bayesianism will be a good way to predict and describe their actions, in addition to providing a normative theory. The overall research program investigating these connections is known as "decision theory", and is a major research topic in economics, psychology, political science, philosophy, and related fields.

2.1 THE DUTCH BOOK ARGUMENT

One historically important argument leading to decision theory starts not with the broad class of all actions, but instead with a very specific class of actions engaged in by very specific agents. In particular, many theorists have focused on the role of degree of belief in explaining gambling behavior to understand its role in guiding behavior more generally. In later sections, I will discuss more general types of action to make sure that the theory is not too distorted by looking at an unusual type of action. But it is useful to start with gambling because of the simplifying features it has.

In ordinary actions, like going somewhere for lunch, there are many different ways things could turn out that matter for the decision. This place might be open or closed. That place might be crowded or quiet. The food might take longer to prepare than usual. And there are plenty of circumstances in which one's decision could be shaped by one's credence in each of these possibilities. It's hard to state directly how a person's credences in each of these propositions might affect their actions, because they interact in so many ways. So it is helpful to start thinking about actions that only depend on a single proposition. And the best example of an action that depends only on one proposition is a *bet* on that proposition.

2.1.1 Bets as contracts

Ordinarily, if you say "wanna bet?", you're taken to be proposing a gamble where the two parties set a price (say \$50), and one of them pays that price to the other if the proposition in question turns out to be true, while the other pays if the proposition turns out to be false. However, it's useful to reconceptualize this sort of bet in a slightly different way. Instead of an agreement for A to pay B \$50 if p is false, and B to pay A \$50 if p is true, we can think of it as A paying B \$50 right now, in exchange for a contract in which B agrees to pay A \$100 if p turns out to be true. If we imagine the bet being settled quickly by some neutral third party, then there is no significant difference between these two descriptions. There is even a third description in which B first pays A \$50, in exchange for a contract in which A agrees to pay B \$100 if p turns out to be *false*. If we think the contracts will be resolved immediately and there is no possibility of anyone reneging on these contracts, then all three of these are equivalent. (But if there is a possibility that someone does renege, then all of these acts have outcomes that depend on more than one proposition and will thus be ignored for now.)

These second and third characterizations give us a useful way to conceptualize a bet. We can think of a "bet on p" as being a contract to pay someone a particular amount of money if p turns out to be true. A bet is characterized by the amount of the payment (the "stakes") and the proposition that determines whether the payment is made. We are usually interested in identifying what prices someone is willing to buy or sell a bet at. The second characterization of the bet above is one in which A is willing to buy a \$100 bet on p for \$50 (and B is willing to sell it at that price), while the third characterization is one in which A is willing to *sell* a \$100 bet on $\bar{p}$ for \$50 (and B is willing to buy it at that price).

2.1.2 Fair prices for bets as a measure of degree of belief

Once we characterize bets as these kinds of contracts, we can think about what it means for someone's buying or selling price to change. If I have high credence that the restaurant is in fact open, I might be happy to buy a bet on this proposition at any price up to \$80. Or if I think the restaurant is probably closed, I might only be willing to pay prices up to \$10 for such a contract. If the price is high enough, I'd be willing to sell the contract. If there is some price such that someone is willing to buy the bet at any lower price, and sell it at any higher price, then that price is said to be that person's "fair" price, and the higher prices are said to be "favorable for selling" and lower prices are said to be "favorable for buying".

This is roughly how betting works at racetracks or sportsbooks—the bookie posts some "odds", which indicate what price they consider fair for a $1 bet on some proposition (e.g., that a particular competitor will win), and they let bettors buy or sell these bets for a particular amount above or below this fair price. (The specific difference they require between the price indicated as fair by their posted odds, and the price they actually buy or sell at, is called the "vigorish", or "vig" for short.) Bettors can buy or sell contracts with any stakes, for proportional prices. The bookie tries to post a price where they expect the total stakes of contracts bought to be close to the total stakes of contracts sold, so that whichever way it turns out, the stakes they win can pay out the stakes they lose, and the vigorish can be their profit. The actual numbers used to represent the odds usually aren't the prices of the contracts themselves, but are related to it in some systematic way. (I will discuss this connection more explicitly in Section 4.1.3.3.) But for now, we will focus on the fair price for a $1 bet on a proposition.

In this section, we are thinking of buying and selling bets on a proposition as the only sort of action that a person's degrees of belief govern. On the pragmatic conception of the doxastic state, the only thing that makes one degree of belief count as higher or lower than another is the difference it makes to these sorts of actions. So if someone has the same fair price for the bets they offer on two distinct propositions, then there is no relevant difference between their degrees of belief in these propositions. If their fair price for a $1 bet on p is higher than their fair price for a $1 bet on q, then at any price in between these values, they would be willing to buy a $1 bet on p, but sell a $1 bet on q. In this case, they are treating p as in some sense "closer to true" than q, so their degree of belief must be higher. Thus, we can arbitrarily choose to use this fair price for a $1 bet on a proposition as the numerical representation of the person's degree of belief in that proposition. Once we use this numerical representation, we can derive some normative constraints on how degrees of belief should behave.

2.1.3 Making a book

A central observation in economics is that if you can find the same product available for different prices in different markets, you can make money by "arbitrage", buying low and selling high. Similarly, if a gambler can find multiple bookies that are offering different odds on the same event, she can often make money by placing different bets with the different bookies. More subtle market players may be able to find ways to extract these guaranteed profits by combining multiple bets. Such a combination of bets that together are guaranteed to return a profit is called a "book" or historically a "Dutch book" presumably because of the early rise of modern markets in Amsterdam.

If you can figure out a way to put together some bets that guarantee a profit, that's generally a good thing. In fact, the word "bookie", for the person who posts odds and buys and sells bets just above and below that price, is short for "bookmaker", because getting the stakes on all the contracts bought to equal the stakes on all the contracts sold is effectively making a book. If there is an imbalance between the stakes of contracts bought and contracts sold on a single proposition, then the bookie is running a risk—if the side with greater stakes ends up winning, then the bookie won't have enough winnings from the contracts on the other side to cover these prizes and will have to eat into their vig (or it may possibly even result in a loss if the difference is greater than the total vig). But this is only a *risk*—there is also a possibility that the side with greater stakes ends up losing, and the bookmaker makes extra profit.

But if the bookie is posting odds for multiple different propositions, they might need to watch out that there is no other book put together against them. The only real goal of the bookie is to profit. But there are several subsidiary goals that emerge from this. Making a book against the bettors is a way to *guarantee* a profit. But it is not essential—even an imbalanced set of bets can be ok, if the bookie gets lucky with which side wins. But one thing the bookie definitely wants to avoid is having a book made *against* them. If they're making decisions just based on the goal of making money in a way that *guarantees* that they will lose money, then they've definitely done a bad job. A skillful bookie would post prices such that this sort of book against themself is impossible.

One thing to note—if two bookies post different prices, a bettor would be able to buy low and sell high and create their own book. Neither bookie sees this as necessarily being a problem. One of them will make money off this bettor, and the other will lose money to this bettor, but if each bookie makes their own books, they'll both turn out ok. More likely, one or both of them will end up with an imbalanced set of contracts. In this case, each of them might think that the other has done a bad job, or they might worry about their own risk, or both. But the one sort of error that each could see up front is if a book can be made against them individually.

2.1.4 Probabilism for bookies

We can derive some constraints on what the prices corresponding to a bookie's posted odds must be like in order to avoid a book being made against them. As suggested in Section 2.1.2, we think of the fair price for a bet on a proposition as operationalizing the bookie's credence in that proposition. In this section, we will assume there is no vig or that it is vanishingly small compared to the stakes of the bets. Thus, whatever the credence is, they must be willing to sell a \$1 contract for any amount greater than this credence and buy one for any amount below it. Furthermore, bettors could buy or sell any combination of bets at these prices.

The consequences are the following. First, the bookie's credence in any proposition should be non-negative. Since the contract involves a possibility of the bookie paying out money, they shouldn't be willing to sell the bet for a negative price—that would amount to *paying* the bettor in order to take on this potential liability! This requirement corresponds to the probabilistic axiom of non-negativity. Second, if the bookie is absolutely certain that a proposition is true, then their price for a bet on that proposition should be equal to the payoff of the bet—buying at a higher price than the payoff guarantees a loss, and since the bookie considers the proposition to be certain, selling at a lower price than the payoff would also guarantee a loss. So the fair price for a \$1 bet would be \$1 and thus, the credence in this proposition must be 1. This corresponds to the probabilistic axiom of normalization.

Third, if the bookie is absolutely certain that p and q aren't both true, then their credence for "p or q" should be equal to the sum of their credences for p and for q individually. For instance, consider the proposition "either Seabiscuit will win or Secretariat will", in a case where we are sure that it is not possible for both horses to win. As far as the bookie is concerned, a \$1 contract for this disjunctive proposition is equivalent to two separate contracts—a \$1 contract that Seabiscuit will win and a \$1 contract that Secretariat will. So if the price for the disjunction is different from the sum of the prices on the disjuncts, then a book could be made against the bookie. If the price for the disjunctive contract is higher than the sum of the prices for the two individual contracts, then two bettors could buy the individual contracts while another bettor sells the disjunctive one, and the result would be a book. Or if the price for the disjunctive contract is lower than the sum of the prices for the two individual contracts, then two bettors could buy the individual contracts while another sells the disjunctive one. In order to avoid both types of books, the bookie needs the price for a disjunctive contract (where there is no possibility of both parts being true) to equal the sum of the prices for the individual propositions. This corresponds to the probabilistic axiom of additivity.

In fact, as proven explicitly by Kemeny (1955) (and in a more abstract mathematical form by de Finetti (1937)), as long as a bookie's degrees of belief satisfy the probability axioms, then there is no finite set of bets that they will accept that constitute a book against them. (For infinite collections, see my discussion in Easwaran (2013b).) So, satisfying the probability axioms is necessary and sufficient to avoid any possibility of having a book made against them. Interestingly, there is no need to assume there is any internal state of the bookie that is their "degree of belief" or even an "attitude"—all that matters is that the bookie posts odds and these posted odds suffice to determine the quantity we end up calling their "degree of belief".

2.1.5 Limitations of this argument

While this argument establishes that bookies should have degrees of belief that satisfy the probability axioms, it has less obvious import for anyone else. For one thing, it relies on the fact that a bookie posts prices, and will sell bets for any stakes at prices above that, and buy bets for any stakes at prices below that, including arbitrary combinations of buying and selling at these prices. Furthermore, it makes the assumption that monetary gain and loss is the only thing that matters to the agent—real bookies operate under legal requirements and often care about things like managing personal relationships, in addition to money. Finally, as mentioned from the outset, this argument is based on the idea that prices of contracts that each only depend on a single proposition are sufficient to characterize an agent's degrees of belief. In actuality, the pragmatic role of degree of belief is to govern *all* sorts of action, not just these.

However, this sort of simplified setting is a useful way to start thinking about pragmatic arguments for probabilism. Many of the issues raised here will help explain why later arguments have been developed the way they are. Readers interested in more detailed discussion of Dutch book arguments should consult Vineberg (2022).

2.2 RAMSEY AND REPRESENTATION THEOREMS

One of the first explicitly philosophical generalizations of the Dutch book argument was given by Ramsey (1926). This argument ended up setting the stage for many more detailed arguments that have followed over the century since then. Ramsey begins by arguing against frequentist and logical interpretations of probability and in favor of what would eventually come to be called the Bayesian interpretation of probability as degree of belief. Before moving to his main argument, he considers the idea that degree of belief can be understood by some kind of phenomenological introspection of the strength of one's beliefs.

> This view would be very inconvenient, for it is not easy to ascribe numbers to the intensities of feelings; but apart from this it seems to me observably false, for the beliefs which we hold most strongly are often accompanied by practically no feeling at all; no one feels strongly about things he takes for granted.

Instead, he argues that "degree of belief is a causal property of it, which we can express vaguely as the extent to which we are prepared to act on it." This leads directly to understanding the doxastic state in terms of its pragmatic role. But he makes some important advances over the Dutch book argument.

2.2.1 Money as a bad measure of value

The core issue that Ramsey addresses is the assumption that money is the way to measure what an agent values. First, it's clear that there are many things that people value that can't simply be acquired with money. The obvious examples are things like friendship and respect. But even in terms of material goods, we can see plenty of examples by looking at the history of technology. In much of the world today, even relatively poor people have access to goods, like basic vaccinations and heated and air conditioned spaces, that even the richest people in the world had no access to a few centuries ago. But even beyond all the things people value that can't be directly translated into monetary terms, it's clear that the value of money is not always the same.

For a relatively poor person, a few hundred dollars might mean the difference between keeping the air conditioning on at home or having it shut off and needing to go to the public library for some relief on hot days, while for a rich person, a few hundred dollars might only make the difference between one excellent dinner and a slightly better dinner. This observation is often described as the "diminishing marginal value" of money—at the margin, each additional dollar generally provides less value than the previous dollars. While there are some standardized economic models of how this works, no model will capture the entire value of money in a general way, because the marginal value doesn't always diminish. As a classic example, someone stranded in another town without access to a credit card or phone might need $50 to get a ticket home. For that person, the first $49 they can put together will have some value (they might mean the difference between a meal tonight and going hungry), but that 50th dollar will be much more valuable than the previous ones, because it suddenly unlocks a new possibility. Because of the way that the value of money depends on other circumstances, it's clear that a bet really isn't an act whose value depends only on one proposition—the number of dollars you get may only depend on one proposition, but the value of those dollars may depend on others (like whether you find an alternate route home, or whether you inherit a fortune, or whether the US economy undergoes massive inflation or deflation).

2.2.2 The outline of Ramsey's argument

Ramsey skips money entirely as a denomination of a person's desires, and instead proposes a general idea of how to define a numerical scale. Just as we precisify the doxastic state in terms of degree of belief and measure it numerically, Ramsey proposes to precisify the bouletic state in terms of what we might call "utility", and represent it numerically. Once this numerical representation is available, Ramsey then proceeds roughly similarly to the Dutch book argument—he still primarily focuses on actions whose outcome only depends on one proposition and uses the numerical utilities of the outcomes to construct a representation of the agent's numerical degrees of belief. But instead of these being explicit monetary bets which can be of any scale he imagines agents faced with a more unrealistic sort of choice. He does this in several steps.

First, he replaces money as the object of value, or even any sort of tangible physical goods that can be collected and added, with "possible worlds". The idea is that agents might want many sorts of things—a person might want to live in a nice house or might want to experience love and friendship with a variety of companions; an animal might want food or might want to reproduce; a non-profit organization might want to increase the number of women who get graduate degrees or might want to eliminate malaria; an artificial intelligence might want to predict the next word correctly or might want to maximize the number of paperclips available; and who knows what an alien or angel might want. But whatever it is that the agent wants, a full specification of a possible world in all its detail will be sufficient to settle everything that the agent cares about.

At this point, Ramsey says,

> If then we had the power of the Almighty, and could persuade our subject of our power, we could, by offering him options, discover how he placed in order of merit all possible courses of the world. In this way all possible worlds would be put in an order of value, but we should have no definite way of representing them by numbers.

The idea here is that whatever it is the subject cares about, and whatever interactions it has with anything else, there should be some totality that summarizes it all, and we can identify which overall outcomes are better or worse than others for the subject. Then, he says,

> We could test his degree of belief in different propositions by making him offers of the following kind. Would you rather have world α in any event; or world β if p is true, and world γ if p is false?

Ramsey will then use these binary choices both to determine numerical representations of the utilities of these worlds *and* to identify the subject's degrees of belief in the proposition p.

Once we have the numerical utilities $U(\alpha)$, $U(\beta)$, and $U(\gamma)$, we can think of these binary choices like the bets the bookie was willing to buy or sell above. Let's assume that $U(\beta) > U(\alpha) > U(\gamma)$. (The choice is generally only interesting when α is between β and γ, and we can rename β and γ if they are in the opposite order.) In that case, choosing the gamble between β and γ is like starting at α, and then paying $U(\alpha) - U(\gamma)$ for a contract on p with stakes $U(\beta) - U(\gamma)$.[1] It's also like starting

[1]There are some metaphysical complications around p itself being true or false in α, β, or γ. Ramsey adopts a kind of logical atomism that allows for this one proposition to be separable from the rest of the features of the world. See Klement (2025) for more. There are additional worries if the truth or falsity of p itself makes a difference to how good or bad the agent takes α, β, or γ to be. Ramsey has some moves to address these worries but we will skip them, since they don't matter for later versions of the argument.

at α and then selling a contract on $\bar{p}$ with stakes $U(\gamma) - U(\beta)$ for $U(\beta) - U(\alpha)$. Because this is a forced choice between two options, and because these worlds can be chosen with any values whatsoever, the axioms on preference that Ramsey assumes in his argument will ensure that an agent will act like a bookie in the original version, being willing to buy up to a particular price, and being willing to sell beyond that. Given the rest of Ramsey's argument in generating numerical utilities (which we will get to shortly), these values will automatically satisfy the probability axioms—there is no need to consider combinations of bets to ensure additivity.

2.2.3 Ethically neutral propositions

Ramsey first limits consideration to "ethically neutral" propositions p—that is, to propositions where changing just their truth value (and nothing else) makes no difference to how you care about any possible outcome of the world. We can imagine here a proposition about the result of a coin flip—even if you care about something that followed from the coin flip, if we imagine *just* changing the outcome of the flip while keeping your wins and losses, you probably wouldn't care about the flip itself. (Maybe you think it would be aesthetically nicer for a particular coin to land heads. In that case, the coin flip is not ethically neutral for you—his assumption is just that there is *some* proposition for which you really don't care *at all* whether it is true or false, holding everything else fixed.)

For an ethically neutral proposition p, Ramsey states what it means for an agent to have degree of belief 1/2 in it. In particular, it means that there are some β and γ that are distinct such that the person is indifferent between the gamble between β if p and γ otherwise and the gamble between γ if p and β otherwise. Ramsey's Axiom 1 states that if a proposition satisfies this property for an agent for some pair β, γ, then it does for every such pair, and furthermore that there is at least one such proposition, which we can call an ethically neutral proposition with degree of belief 1/2.

2.2.4 Measuring utility steps

Given an ethically neutral proposition of probability 1/2, Ramsey then defines what it means for the difference in utility between α and β to be equal to the difference in utility between γ and δ. He defines this concept as holding when a person is indifferent between the gamble between α

and δ on an ethically neutral proposition of probability 1/2, and the gamble between β and γ on that same proposition.

Ramsey adds four more axioms that ensure that this concept of "equal difference" really behaves the way we expect it to. His Axiom 3 states that if an agent is indifferent between two gambles and is also indifferent between the second and a third, then the agent is indifferent between the first and the third. His Axiom 2 states that when the person is indifferent between the gamble between α and δ and the gamble between β and γ on one ethically neutral proposition of degree of belief 1/2, then the person is indifferent between these gambles on all such propositions.

Ramsey's Axiom 2a states that if the person finds α to be better than β, then they must find γ to be better than δ. This is reasonable because worsening the outcome of α to β can only result in indifference if the other outcome is improved, so γ must be better than δ. Because of the symmetry of the setup, this also ensures that if the person finds β to be better than α, then they must find δ to be better than γ. And since he has implicitly assumed that the agent has preferences among all options, this means that if they are indifferent between α and β, then they must be indifferent between γ and δ. This axiom ends up ensuring that the numerical degrees of belief calculated by the method in Section 2.2.2 will always be non-negative, and normalized.

But there is also a more complicated Axiom 4 that is necessary. In order to ensure that this relation of "equal differences" really is an equivalence relation, he needs to assume that if an agent is indifferent between the gamble between α and δ and the gamble between β and γ, and they are also indifferent between the gamble between γ and ζ and the gamble between δ and η, then the agent should be indifferent between the gamble between α and ζ and the gamble between β and η. With some detailed checking of the conditions, we can see that this does make sense if a person really does have numerically-measurable utilities for these outcomes. But this axiom is harder to motivate directly. Ramsey does not explicitly spell out the argument that numerical degrees of belief will satisfy additivity (and I won't either), but this axiom plays an important role in that argument.

2.2.5 Using utility steps to define numerical utility

Finally, with the concept of "equal differences" defined, Ramsey uses this to define a numerical representation of the utility of every possible world. He uses four more axioms to do this. Two of these axioms are simple assumptions about the existence of worlds with particular utility differences compared to others. Axiom 5 states that for any α, β, γ, there is a world such that the step from α to β is equal to the step from γ to δ—among other things, this ensures that there is no best or worst possible world. It ensures that whatever the difference is between α and β, there is an infinite sequence of worlds, both upward and downward from γ, each consecutive pair of which have precisely the same difference as that between α and β. Axiom 6 states that for any α, β, there is a γ such that the step from α to γ is equal to the step from γ to β.

The other two are not stated explicitly, but are named with standard mathematical concepts. Axiom 7 he states as "Axiom of continuity:— Any progression has a limit." By a "progression", I assume that he means something like a "Cauchy sequence"—this is an infinite sequence of worlds, such that for every difference, all but finitely many pairs of elements of the sequence have less than that difference from each other. (Ramsey hasn't explicitly defined what it is for the difference between one pair of elements to be less than the difference between another pair, but it is plausible that the previous axioms make this concept behave how we expect.) For a progression like this, a "limit" is a single value such that for every difference, all but finitely many elements of the sequence have less than that difference from the limit. Regardless of what precisely he means by this axiom, a version of this is one part of the characterization of the real numbers.

Axiom 8 he calls the "Axiom of Archimedes" and gives no definition, but there is only one natural interpretation of this. Given any four possible worlds $\alpha, \beta, \gamma, \delta$, with β better than α and δ better than γ, the sequence of worlds starting with γ, whose adjacent worlds have the same difference as that from α to β, must have some finite number of steps after which it passes δ.

These four axioms ensure that the worlds can be represented as having numerical utilities in one-to-one correspondence with the real numbers. Pick any arbitrary pair of worlds with α better than β, and label β with the number 0 and α with the number 1. The first of these four axioms ensures that there is an infinite sequence of worlds whose utilities correspond to the integers. The Axiom of Archimedes ensures that every world either has utility corresponding to one of these integers or in between some pair of them—no world has utility greater than every positive integer or less than every negative integer. The second of the four axioms ensures that we can subdivide these integer steps by every power of 2 and find worlds with each such utility. For any real number, the Axiom of Continuity for worlds ensures that there is a world with that real number utility. And for every world, the Axiom of Continuity for the real numbers ensures that there is a real number that represents the utility of the world.

The numerical scale for value behaves like the Celsius or Fahrenheit numerical scales for temperature. There's something arbitrary about using either scale, just like there was something arbitrary about the choice of α and β as 1 and 0. Celsius and Fahrenheit disagree about where zero falls, and what the unit is, but (as I will discuss in greater detail in Chapter 4) they represent the same underlying facts. But any two numerical scales that agree about "equal difference in value" facts will be like these—the only difference is the arbitrary choice of where you put 0 and what the unit is. Furthermore, for any such scale, the numerical degrees of belief calculated as described in Section 2.2.2 will be the same.

2.2.6 Conclusion of Ramsey

With this numerical scale for value, Ramsey can then define degree of belief just like we did in the Dutch book argument—if an agent is indifferent between one guaranteed numerical value and a gamble on a proposition between two different numerical values, then as mentioned in Section 2.2.2, we can calculate the "price" of a "bet" and derive a numerical representation of the agent's degree of belief. The axioms 1, 2, 2a, 3, 4, 5, 6, 7, 8 that Ramsey uses to ensure that utilities can be represented on a numerical scale also ensure that the price of a bet on a proposition calculated with one set of options the agent might consider is always the same as the price of a bet on that same proposition calculated with a different set of options.

Of course, in this argument, the devil is in the details of the assumptions about rational indifference that Ramsey makes to ensure that "equal difference in value" behaves the way that we expect it to, and that there is a correspondence between worlds and real numbers. His assumptions are clearly mathematical in nature, with no obvious epistemological or psychological justification for them. But they make sense for any sort of doxastic subject that has preferences among outcomes, and preferences among uncertain gambles, not just people acting as bookies. There is no need for the subject themself to have internal states that correspond to utilities and degrees of belief—all that matters is that they have preferences that obey Ramsey's axioms. The representation he constructs shows that they are acting *as if* they have utilities and degrees of belief, and this is all that matters. Once Ramsey showed that the argument could be abstracted away from the particulars of money and gambling, other economists, psychologists, statisticians, and philosophers developed more realistic versions of this argument.

2.3 SAVAGE'S REPRESENTATION THEOREM

Leonard "Jimmie" Savage, in his (1954) developed a representation theorem of a more modern form. Like Ramsey, Savage assumes a set of axioms on rational preference and uses them to derive a numerical representation of degrees of belief that satisfy the probability axioms. But rather than considering Ramsey's particular class of artificial gambles, offering complete states of the world as prizes for individual propositions being true and false, Savage begins with a reconsideration of the objects of choice. Ramsey's logical atomism led him into difficulties, because propositions are parts of worlds, which are the objects of value, and he introduces this strange fiction that someone "has the power of the Almighty" to bring about the actuality of particular worlds (depending on the truth of certain propositions).

2.3.1 Consequences, acts, states, events

Savage instead uses the word "consequences" for whatever it is that the agent values. These might be full possible worlds as Ramsey suggests, but for most agents they will be something somewhat more restricted (and at worst, they can be identified with equivalence classes of possible worlds). Savage notes that for an agent acting in the world, there are only some parts of the world that are under the agent's control, which he calls the "act". Whatever else, together with the act, plays a role in determining the consequence, he calls the "state". The important thing for him is that the state is completely outside the agent's control, while the act is completely inside the agent's control so the act and the state are independent of one another. (This is an assumption that will be questioned in Section 2.5.) Any set of states is called an "event".

Strictly speaking, for Savage, *consequences* are the only things for which the agent has utilities, while *events* are the only things for which the agent has degrees of belief. For *acts*, the agent has preferences. Preferences among acts are the things that the agent needs to have in some manifest way. Utilities and degrees of belief may or may not be real—but for any agent that satisfies Savage's axioms, they can be constructed, and furthermore, the representation theorem shows how to represent them with real numbers. Notably, the agent does not have degrees of belief about their own acts—just about events, which are composed of the states outside their control. This idea has been challenged by Hájek (2016), but defended by Liu and Price (2020), who argue that a version of this idea is also present in Ramsey. It is notable that these states and events are somewhat different from the kinds of possibilities and propositions that I considered in Section 1.3.

While acts are initially identified with parts of the world that are under the agent's control, it is important for Savage to consider the agent's preferences not just over actual acts that they can do but over hypothetical acts as well (akin to Ramsey's gambles with the Almighty). But where Ramsey considers only binary choices between pairs of options, each of which is either one world for certain or a gamble between two worlds that depends on a single proposition, Savage considers *all* functions from states to consequences as hypothetical acts. Even though only some functions assigning consequences to each state are realized by possible physical acts, Savage's axiom P1 states that the agent has (perhaps only implicitly) a linear ordering of preferences (i.e., the agent's preferences form a symmetric, transitive, anti-symmetric, and total relation) over *all* pairs of such functions, and not just the ones that the agent could actually encounter as a result of feasible actions. (Some of the role of the infeasible acts is defended by Dreier (1996).)

2.3.2 Orderings on consequences and events

Given this overall preference ordering, Savage first considers the hypothetical acts that result in the same consequence regardless of the state of the world. Savage's axiom P5 states that there is at least one such preference that is strict (so that there are at least two distinct consequences). Preference among these acts (which is a linear ordering, by P1) defines an ordering on the consequences, which is the first step toward constructing utilities. He also gives an axiom P7, stating that if f and g are two acts, and if for every consequence of f, the constant act always yielding this consequence is preferred to the act g, then f must be preferred to g (and conversely, if g is preferred to every one of these constant acts, then g is preferred to f), ensuring that if one act always yields a better consequence than another, then it is preferred.

Then he considers the "simple gambles", hypothetical acts that would guarantee one consequence on the states in one event, and a second consequence on all states not in that event. These two consequences are called the "prizes"—whichever consequence has higher utility is the "good" prize and the other is the "bad" prize. Savage defines an ordering on events by saying that E is above E' if the agent prefers the gamble with a given set of prizes and the good prize on E to the gamble with the same prizes and the good prize on E'. Savage's axiom P4 states that these preferences are consistent—if the agent prefers the gamble with a given good and bad prize based on E rather than E', then the agent prefers the gamble based on E rather than E' with *any* good and bad prizes. Since this ordering on events is derived from preference on acts, P1 ensures that it is a linear ordering. This is the first step toward constructing degrees of belief.

2.3.3 Conditional preference

Savage makes one further significant definition involving implicit *conditional* preference among acts. Consider the four tickets shown in Table 2.1: ticket A gets you admission to the new water park if the weather is nice, and gives you free admission to the art museum otherwise; ticket B gets you admission to the new water park if the weather is nice, and gives you free admission to the science museum otherwise; ticket C gets you admission to the football game if the weather is nice, and gives you free admission to the art museum otherwise; and D gets you admission to the football game if the weather is nice, and gives you free admission to the science museum otherwise (Table 2.1).

If you've never been to the water park, art museum, or science museum, then you don't know how much you'll enjoy any of these activities—the outcome for you will depend not just on which ticket you pick, and what the weather is like, but all sorts of other fine-grained states of the world. (Maybe the water slides will be exciting and maybe

TABLE 2.1 Ticket options and their corresponding activities

Ticket	If weather is nice	Otherwise
A	Water park	Art museum
B	Water park	Science museum
C	Football game	Art museum
D	Football game	Science museum

they won't; maybe the art museum contains something that will change your life or maybe it's all just knockoffs of Monet; maybe the science museum will help you really understand your local weather patterns, or maybe it'll just be boring exhibits.) The same is clearly true for the football game—your enjoyment will depend on the details of how the game goes.

However, it seems clear that you don't need to consider all the possible consequences in all those states to compare certain pairs of tickets. Since tickets A and B give you access to the water park if the weather is nice, it seems clear that comparing them only depends on the consequences of going to the museums when the weather isn't nice (possibly including whether they have leaky roofs). Since the first and third both give you access to the art museum if the weather isn't nice, comparing them only depends on the states of the water park and football game in which the weather is nice. If you prefer the first to the second, then you should prefer the third to the fourth; if you prefer the first to the third, then you should prefer the second to the fourth.

Savage formulates this principle more abstractly. Let f, f', g, g' be four acts (considered as functions from states to consequences) and E be an event. Let a be the act that agrees with f on states inside E and with g on all other states. Let b be the act that agrees with f inside E and with g' on all other states. Let c be the act that agrees with f' inside E and with g on all other states. Let d be the act that agrees with g' on all other states. Savage uses preferences among a, b, c, d to define *conditional* preferences among f, g, f', g'. That is, if an agent prefers a to c, this means they prefer f to f' conditional on E (since the differences between a and c just are the differences between f and f' in E states), and if the agent prefers a to b, this means they prefer g to g' conditional on $\bar{E}$ (since the differences between a and b just are the differences between g and g' in $\bar{E}$ states).

Savage's axiom P3 is that these conditional preferences are consistent. That is, an agent that prefers a to b, must prefer c to d, and an agent that prefers a to c must prefer b to d. In general, it means that whatever preference an agent has between two acts must be driven entirely by the consequences in states where they differ—if there are states where these acts have the same consequences, and both acts are modified in the same way on those states, then the preference among the resulting acts must be the same as among the original acts. This assumption, which enables him to talk about conditional preferences in a meaningful way, is probably the most complicated one of Savage's assumptions, and it has attracted the most attention. It is often referred to as the "sure thing principle", though Savage used this term more informally for some of the ideas that motivated this assumption and others.

Savage also uses conditional preference to formalize the idea of the differences between states and consequences in his axiom P2—it states that if consequence x has higher utility than consequence y, then the agent prefers x to y conditional on any event. That is, the utility for a consequence doesn't depend on what state it occurs in—everything that the agent cares about should be thought of as part of the consequence rather than part of the state.

2.3.4 Savage's construction of numerical representations

With conditional preference and these orderings on utility and degree of belief, Savage is able to get quite a lot. A set of events such that every state is contained in exactly one of these events is called a "partition". If there exists a partition containing exactly n events, and each of these events has the same degree of belief, then Savage represents this degree of belief with the number $1/n$. The one axiom of Savage's that hasn't been mentioned is his P6, which ensures that such partitions exist for every n. The existence and consistency of conditional preference ensures that the numbers assigned in this way are consistent and satisfy additivity. (It is clear that they satisfy non-negativity and normalization.)

For instance, the fact that $E_1, \ldots, E_6$ form a partition with equal degree of belief is constituted by the fact that the agent is indifferent between all pairs of gambles on these individual events, like Acts a and b, illustrated in Figure 2.1. But this indifference is also part of the fact that the agent is indifferent between these two acts conditional on the event of $E_3 \cup E_4$. This conditional indifference means that the agent must also be indifferent between Acts c and d, illustrated in Figure 2.2.

By putting these sorts of comparisons together, we can see that an agent who is indifferent between gambles on each pair of elements in a fine partition into six events must also be indifferent between gambles on each pair of events consisting of unions of the same number of these events,—including the pairs (which can be used to form partitions into three events) and triples (which can be used to form partitions into two events). The fact that the number $1/6$ is assigned as the representation of the degree of belief for each of the members of the fine partition, and $1/3$ or $1/2$ for the coarser partitions, corresponds to the fact that $1/3 = 2/6$ and $1/2 = 3/6$.

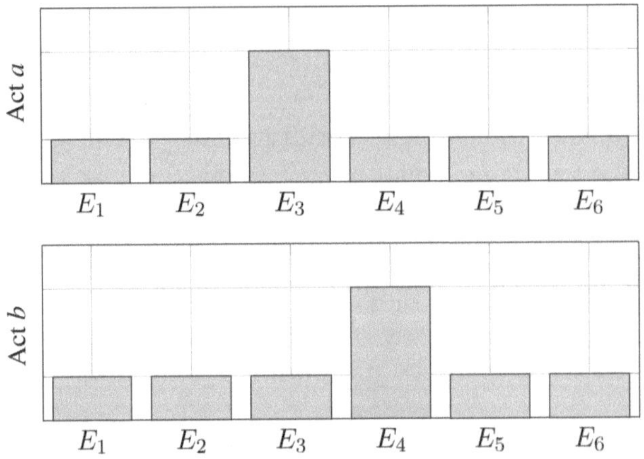

Figure 2.1 Having the same degree of belief in each of the events $E_1, \ldots, E_6$ means being indifferent between events like a and b.

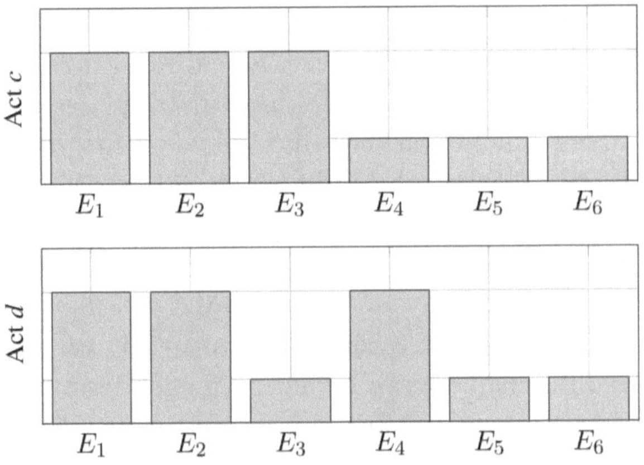

Figure 2.2 Since acts c and d yield the same consequences as a and b respectively in the event $E_3 \cup E_4$, and the same consequences as each other in all states outside $E_3 \cup E_4$, axiom P3 means that an agent indifferent between a and b must be indifferent between c and d as well.

We can use these partitions to numerically represent the degree of belief for any event whose degree of belief is a ratio of two integers. We can then use the linear ordering on degrees of belief to assign real number representations to all events. This works because every real number is defined uniquely by the set of rational numbers that it is greater than, and the set of rational numbers that it is less than. (In fact, this is the key to our familiar decimal notation system for real numbers. The fact that we can write π as 3.141... just means that $3 < \pi < 4$ and $31/10 < \pi < 32/10$ and $314/100 < \pi < 315/100$ and $3141/1000 < \pi < 3142/1000$ and so on.)

Finally, we can derive numerical representations of utility from these numerical representations of degree of belief. Again, we arbitrarily represent the utility of some consequence as 0 and some higher utility as 1. We can identify which consequence should be represented with utility -1, because it is the one where the agent is indifferent between certainty of 0 and a gamble between this consequence and the consequence of utility 1 on an event with degree of belief 0.5. We can identify which consequence's utility should be represented with x between 0 and 1 by seeing which certainty leaves the agent indifferent between it and a gamble on an event with degree of belief x between 1 and 0, and similarly for x between 0 and -1. And we can identify which consequence's utility should be represented with $x > 1$ (and similarly $x < -1$), because a gamble between that consequence and 0 on an event of probability $1/x$ should be equivalent to certainty of 1. Thus, while Ramsey defined value first and probability later, Savage reverses the order, by considering the distinction between acts, states, and consequences. Savage's axioms P1, P2, P3, P4, P5, P6, and P7 are somewhat more directly connected to an intuitive understanding of preference than Ramsey's axioms.

2.3.5 Expected value by summation

While the Dutch book argument and Ramsey's representation theorem only considered acts with one or two possible outcomes, Savage's representation theorem applies much more generally. Consider an act f for which there are only finitely many possible consequences. For each value v, consider the set of states in which act f produces a consequence with that value. Let $P(f = v)$ be the numerical degree of belief in this event. If we multiply these together and then add them up over all of the finitely many values, then we can define

$$V(f) = \sum_{v} v \cdot P(f = v).$$

It turns out that Savage's axioms, together with his definitions of numerical degree of belief and value, entail that act f is preferred to act g if and only if $V(f) > V(g)$.

In the special case where f is guaranteed to have value v and g is guaranteed to either produce the value 0 or the value 1, then we can see that the person will be indifferent between f and g if and only if $v = P(g = 1)$. That is, the prices of simple bets that we used in section 2.1 and that Ramsey used in section 2.2 are special cases of the more general rule relating numerical degrees of beliefs and values to preference.

When there are finitely many possible values of consequence that an act could result in, the probability axioms ensure that the numbers $P(f = v)$ will sum to 1. If each consequence is equally likely, then $V(f)$ will be the average of these values. But even if not, we can think of it as a sort of "weighted average" of the possible values of the consequence, where the weighting of each possible value is given by the credence that one has in that consequence occurring. In official terminology, $V(f)$ is called the "expected value" of act f, and if we replace v in the calculation by any other assignment of numbers to possible states of the world, then the formula gives what is called the "expectation" of that quantity. I will sometimes write the expectation of quantity q as "$Exp(q)$", sometimes with a subscript indicating what probability function is used to calculate the expectation.

This terminology is slightly misleading, since we don't necessarily expect that the quantity will actually equal this number. If a pediatrician meets a parent and is uncertain how many children the parent has, it may be that the "expected number of children" is 1.89, even though the pediatrician knows for certain that the parent either has 1, or 2, or 3, or ... children, and doesn't have any fractional children. No particular family is just like the average family, and no particular possible value of an act needs to be just like the expected value of the act, which is a kind of average of its possible values. However, preference among acts corresponds to this expected value, in cases where there are only finitely many possible outcomes.

One mathematically important feature of expected value is that, when some numerical quantity is a sum of two others, the expected value of the sum will always be exactly equal to the sum of the expected values of the others. That is,

$$Exp(x + y) = Exp(x) + Exp(y).$$

For instance, if we assume that the total number of children in every family is the total number of boys plus the total number of girls, then we can show that the expected number of children is equal to the expected number of boys plus the expected number of girls. This is because the probability of having precisely one child is the probability of having precisely one boy and no girls plus the probability of having precisely one girl and no boys, and the probability of having precisely two children is the probability of having precisely two boys and no girls plus the probability of having precisely one boy and one girl plus the probability of having precisely two girls and no boys, and so on. Similarly, the probability of having precisely one boy is the probability of having precisely one boy and no girls plus the probability of having precisely one boy and one girl plus the probability of having precisely one boy and two girls and so on. With a bit of algebraic work, one can show that every term in the calculation of $Exp(x + y)$ shows up precisely once in the calculation of $Exp(x) + Exp(y)$, and vice versa, so they must be equal. This feature is known as the "linearity of expectations", and it applies when expected value is computed with respect to any probability function, not just degrees of belief.

Although Savage requires every function assigning consequences to states to count as an act and thus includes these bets that assign only finitely many values, his framework allows for more realistic complexity as well. Given the infinitely many ways we are each uncertain about the details of the world, any real-world act that one might perform is going to have infinitely many subtle differences in the way that it could turn out that will matter for how good or bad the outcome would be. In deciding between two restaurants for lunch with a visiting friend, one doesn't just consider how far the walk is and how much one likes the food. One is also concerned for how much one's friend might like the food at either place, the various possibilities for delays that could arise at either restaurant, the subtle differences in how the food might be compared to yesterday, and so on. Although one might not consciously consider all of these possibilities (and certainly not in all their infinite variety), any of it can play a role in one's decision.

2.3.6 Expected value by integration

If there are infinitely many different consequences that might result from an act, then the summation formula mentioned above can't generally be used to directly compute the expected value of the act. However, in many cases it is still possible to *indirectly* compute the expected value of the act. We say that one act *dominates* another if, in every state, the consequence of the first act is better than the consequence of the second act. Savage's axioms ensure that if one act dominates another according to how an agent values consequences, then the agent prefers the dominating act.

Even if an act has infinitely many possible outcomes, so that the summation formula for expected value is undefined, it will often dominate some acts with only finitely many consequences and be dominated by others (though of course there are also some that neither dominate nor are dominated by it). We can prove that the expected value of any act that dominates it will be strictly higher than the expected value of any act that it dominates. As long as there are some upper and lower bounds to the utility of the possible consequences of this act, we can in fact find acts that it dominates and acts that dominate it whose expected values are arbitrarily close together. Thus, there will always be precisely one real number that is above the expected values of all the acts that it dominates, and below the expected values of all the acts that dominate it, and we can define the expected value of the act to be this real number. This is one way that the concept of an "integral" in calculus is defined.

2.3.7 Practical approximation of expected value

One important thing to keep in mind: although Savage's axioms ensure that the degree of belief and the utility functions exist, and that preferences among acts must correspond exactly to comparisons of expected values, Savage doesn't say anything about *how* the agent identifies which acts they prefer. In particular, he does not say anything that suggests that rational agents should try to calculate these expected values in order to figure out what to do! For one thing, there's no reason to think that a rational agent would have considered Savage's particular way of using numbers to represent degrees of belief and degrees of value. For an artificial intelligence or a group of humans, it could well be that different preferences are identified by different subsystems or members—there's no reason to think that the full preferences or anything sufficient to calculate them are stored in one place.

What Savage's result shows is that in order for an agent to be consistent and rational in their preferences among acts, they must have the preferences among acts that a perfect calculator equipped with their credence and utility functions would calculate by means of this formula. (The credence function is unique and the utility function is unique up to choice of where to put 0 and what size to make the unit.) This is much like Newton's result showing that in order for someone to successfully catch a ball, they must put their hand in the place that would be recommended by a perfect calculator equipped with the initial velocity of the ball and the equations of motion. (The numbers involved are unique up to choice of what size to make the unit—feet or meters or something else.)

But obviously, even if people don't do it perfectly, we can do a pretty decent job at forming preferences that roughly satisfy Savage's axioms or placing our hands to catch balls roughly where Newton's equations predict they'll be without performing these calculations. If one wants to design a way for corporations to make decisions, or a way for robots to catch balls, it might turn out that having someone explicitly run the calculations is a step in one relatively effective process for these goals. Humans catch balls in a different way (McLeod and Dienes, 1996), and we probably form preferences in some other way that doesn't precisely satisfy Savage's axioms. But Sven Neth (2024) has shown that approximately satisfying Savage's axioms will enable an approximate representation with real numbers. What matters is doing better at approximating these axioms, however one manages to do it. Some specialized agents might have explicit representations of utilities and degrees of belief and perform these calculations, just as SpaceX and NASA do explicit calculations using physical laws when trying to catch high-velocity objects coming from space. But for the rest of us, there's no more reason to do the expected utility calculations than there is for outfielders to study Newton's laws of motion.

2.4 RISK AVERSION

One set of challenges to standard expected value arises from the phenomenon of "risk aversion". This is specifically a challenge to Savage's "sure-thing principle", and his definition of consistent conditional preferences. Consider the following four acts, all dependent on the draw of

TABLE 2.2 Allais's choices

Act	Tickets 1–9	Ticket 10	Tickets 11–100
A	\$5 million	\$0	\$0
B	\$1 million	\$1 million	\$0
C	\$5 million	\$0	\$1 million
D	\$1 million	\$1 million	\$1 million

a random ticket numbered from 1 to 100. (This example is due to the Nobel prize-winning economist Maurice Allais (1953).) On act A, one gets \$5 million if a ticket numbered 1–9 comes up, and nothing otherwise. On act B, one gets \$1 million if a ticket numbered 1–10 comes up, and nothing otherwise. Most people prefer act A to act B, because the chances aren't very different, but the prize is much bigger. On act C, one gets \$5 million if a ticket numbered 1–9 comes up, nothing if ticket 10 comes up, and \$1 million if a ticket 11–100 comes up. On act D, one gets \$1 million no matter which ticket comes up. Here, most people prefer D, because they don't want to run the risk of losing out on a guaranteed million (Table 2.2).

This is a direct contradiction of Savage's "sure-thing principle". (Compare Tables 2.1 and 2.2, specifically with the event of a ticket 11–100 being drawn in place of the weather being nice. There, we said one should prefer A to B if, and only if, one prefers C to D.) No matter what degrees of belief one has for each of the 100 different tickets, and no matter what utilities one has for \$0, \$1 million, and \$5 million, there is no way for the expected value of A to be above B unless that of C is also above D.

The orthodox response is to say that at least one of these preferences is mistaken, probably that for the certainty in D over the risky opportunity in C (unless the value of additional money falls off very quickly above a million). This response is endorsed by Savage in the later editions of his book—he notes that he was tempted by the preference for D over C, but upon thinking about it further, was inclined to construe this initial thought as a mistake. The psychologists Daniel Kahneman and Amos Tversky (1979) used this example as one of the motivations for their "prospect theory", which is meant to be a description of the irrational (as they see it) ways in which ordinary people actually tend to diverge from the norms of Savage's theory. John Quiggin (1982) makes this irrationality more comprehensible by casting it in a mathematical formulation where people pessimistically overweight their degrees of belief that are attached to worse consequences, while underweighting their degrees of belief that are attached to better consequences. On any of these views, people still *ought* to obey Savage's axioms, and if they did, they would have degrees of belief that satisfy the probability axioms, so the probability axioms are still an ideal for decision makers.

However, another response is to say that Allais's case and others like it actually show that Savage's axioms are not requirements of rationality. In recent years, this case has been particularly made by Lara Buchak (2013). She develops a theory that is formally identical to that of Quiggin. However, she argues that this over-weighting of the low-value states and under-weighting of the high-value states is *not* irrational pessimism, but just a different permissible way of caring about one's possible outcomes.

One way to understand this is to think of a decision in light of uncertainty as a *political* decision rather than a solitary one. (Buchak (2017) uses the theory of risk-aversion to motivate this political theory, but here I suggest the reverse.) We imagine each state of the world as really existing in parallel, and a choice of act is a choice to have every self in every state perform this same act. Some of them end up well off as a result, and others end up badly off. The decision in Allais's case to take C is like a decision for a society to undertake a policy that gives 9 people \$5 million, 90 people \$1 million, and 1 person nothing, while the decision to take D is like a decision for a society to undertake a policy that gives all 100 people \$1 million. If it can be legitimate for a society to take the fate of the worst-off members into greater account than the fate of the best-off members, then it can be legitimate for an individual to take the fate of their worst-off possible future selves into greater account than the fate of the best-off possible future selves, rather than treating them each in direct proportion to their credence in that outcome.

The way this works formally is that, in addition to a personal probability function and value function, each decision maker has a personal "risk-weighting function" R that is used to re-scale probabilities. Each possible outcome of an act makes a contribution to the expected value in proportion to its value, but instead of also being in direct proportion to its probability, it depends on whether this value is one of the better or worse possible outcomes of the act. The "risk-weighted expected utility (REU)" of an act that has finitely many possible outcomes is given by:

$$REU(f) = \sum_v v \cdot (R(P(f \geq v)) - R(P(f > v))\,.$$

(This is not the formula Buchak gives, but an equivalent one that I develop in Easwaran (2026).) In order for this calculation to make sense, the function R needs to meet a few conditions. In order for every possible outcome to make a positive contribution to the value, we need R to be an increasing function. In order for an act with a guaranteed outcome to be as valuable as that outcome, we need $R(1) - R(0) = 1$. Thus, Buchak requires that R be an increasing function with $R(0) = 0$ and $R(1) = 1$. (I show in my (2026) that this last condition can be removed, but that R being an increasing function is essential.)

Buchak shows that decision makers whose preference among acts corresponds to REU rather than to V don't satisfy all of Savage's axioms, but do satisfy a very closely related set of axioms on preference. Furthermore, she proves a representation theorem like Savage's. For anyone whose preferences satisfy this slightly modified set of axioms, she shows that there is a unique probability function P and risk-weighting function R, and a value function V that is unique up to choice of 0 and unit, such that the person's preferences among acts exactly correspond to comparisons of REU of those acts. She argues that these modified axioms are a better understanding of the requirements of rationality than Savage's axioms.

There is controversy over whether Buchak or Savage does a better job of capturing the requirements of rationality. However, for the purposes of epistemology, the upshot is the same. Agents are the sort of creatures that have preferences over acts. There are some conditions on preference that are required for rationality. For an agent's preferences to meet these conditions, these preferences must agree with the ones that are generated by a particular set of degrees of belief in various events, and a particular set of utilities for consequences. These degrees of belief satisfy the probability axioms. Therefore, any rational agent must, at least implicitly, have degrees of belief that satisfy the probability axioms.

2.5 ACTS, STATES, AND CAUSATION

For both Savage and Buchak, "acts" are functions that assign "consequences" to various "states". Degree of belief is an assignment of numbers to "events", which are sets of states. However, note that this interpretation means that "acts" are quite a different sort of metaphysical entity from "events" or "states". On this formulation, events are ways that the world might be *independent* of one's action, and events are the only things in which one has degrees of belief. One can't have degrees of belief about which act one will perform. Levi (1993), Liu and Price (2019, 2020), and Fusco (2025) have suggested that this is exactly right—when deliberating about an action, a free agent must be treating the acts as objects of decision that she can *make* true or false, rather than as features of the world that might *be* true or false in which she can have higher or lower degree of belief. But Rabinowicz (2002) and Hájek (2016) think that this sharp epistemic distinction between propositions about events in the world and about one's acts is unsustainable. I might be able to deliberate about whether or not to take a bet, while still thinking it is more likely than not that I will take it.

2.5.1 Evidential decision theory

Thus, Richard Jeffrey (1965) developed a slightly different representation theorem in which acts are just propositions like any other. Each proposition is represented by a set of possibilities in which it is true, regardless of whether it is a proposition about the self or about the world. Preference over propositions is about the "news value" that represents the extent to which learning that it is true would constitute good news, regardless of whether the proposition looks like one of Savage's "acts" that the agent can control, or the "consequences" for which Savage's agents have utilities. Learning that it is raining (a fact about the external world) might be good or bad news, because of the ways that it makes good or bad possibilities more or less likely. Similarly, learning that you will decide to go for a run this afternoon (a fact about your own behavior) might also be good or bad news, again because of the ways that it makes particular good or bad possibilities more or less likely (both by bringing about your endorphin rush and exhaustion, and as evidence about the absence of rain and surprise parties). A proposition that is evidence for a particular good possibility might nevertheless be bad news if it is even stronger evidence for some other bad possibilities. Jeffrey thinks that we shouldn't draw any sharp distinction between propositions about the self and the world, because the self just is part of the world, and we can form judgments about the probability of our own future behavior. Because preference is fundamentally about the evidential value of propositions for Jeffrey, his theory is called "evidential decision theory".

Jeffrey again proves a representation theorem like those of Savage and Buchak, with a different set of assumptions for how a rational agent should process "news value". There is a set of doxastic possibilities, and every proposition is a subset of this set. A set of propositions is called a "partition" if and only if every possibility is in exactly one of these propositions. Jeffrey (using some results by the mathematician Ethan Bolker) shows that for any rational agent satisfying his assumptions, there is a numerical news value function NV such that an agent considers a better news than b if and only if $NV(a) > NV(b)$. Furthermore, there is a probability function P such that for any proposition a and finite partition $\{e_1, \ldots, e_n\}$, the news value function NV is related to P by means of this equation:

$$NV(a) = \sum_{i=1}^{n} NV(e_i \cap a)P(e_i|a),$$

where $P(e_i|a)$ is conditional probability (to be discussed more in Chapter 6), given by $P(e_i|a) = P(e_i \cap a)/P(a)$. In the special case where a is an "act" and the elements of e_i are completely independent of a (so that $P(e_i|a) = P(e_i)$) and give complete specifications of every aspect of the world independent of the agent (so that $e_i \cap a$ is a set consisting of a unique possibility, which Savage would consider a "consequence"), this formula agrees with that of Savage. But it also gives probabilities and news values for propositions other than acts and events.

2.5.2 The Newcomb problem

However, once acts are just like any other propositions in the world, strange situations can arise. Acts and states are no longer conceptually or causally independent. The classic example is the "Newcomb problem" (Nozick, 1969), which arises from considering a case in which a psychologist is able to predict the decision one will make. But I think it is easier to consider a slightly different case that puts some of the issues more starkly. Late in his career, the statistician (and smoker) R.A. Fisher (1959) argued that the evidence available at the time was consistent with the possibility that smoking doesn't *cause* lung cancer, but instead is only *correlated* with it. The idea is that perhaps people with early-stage undetected lung cancer have a sort of itch in their lungs that could be soothed by smoking. Smokers are much more likely to get lung cancer just because people with lung cancer are more likely to take up or continue smoking, and not because smoking causes lung cancer.

Although this theory is clearly incompatible with modern data (worldwide, trends in lung cancer seem to follow trends in smoking, including changes in smoking due to regulatory intervention, and medical researchers have a fairly broad understanding of the pathways through which burned organic matter causes mutations in DNA), we can think about the decision problem facing a potential smoker in a hypothetical universe where Fisher's theory of cancer and smoking was correct. Although it would be bad *news* to learn that one is going to take up smoking (because it would be evidence that one is likely to develop cancer), it would be perfectly *harmless* to do so. In fact, if one gets some slight pleasure or relief from smoking, then the news that it is correlated with cancer seems to provide no reason not to choose to smoke. Although evidence about one's acts can be bad news, it seems like bad policy to choose the acts that are good news—instead one should choose the acts that actually *cause* better consequences. (Note that this situation can't be properly represented in Savage's framework, since the acts and states are not fully independent of each other.)

Another decision problem that is often said to have the same form is the Prisoner's Dilemma (Axelrod, 1980). In this problem, two co-conspirators have been arrested, and prosecutors have enough evidence to convince each of a small crime, but not enough to convict either of a large crime. The co-conspirators can each choose to either defect (giving the prosecutors evidence about a larger crime that the other committed, in exchange for a slight reduction in sentence for any crime they are convicted of) or cooperate (not aiding the investigation). For each prisoner, defecting makes things a lot worse for the other, but a little better for themself. Assuming these prisoners will never interact with each other again, but know that they have similar character traits and are likely to behave similarly, defecting in this situation is like smoking in the above situation—it *makes* things slightly better (by easing one's own sentence), but it is *evidence* that things will get a lot worse (because it is evidence that the other prisoner also defected, which makes it likely one will be convicted of the larger crime).

Savage's decision theory can't properly express either of these problems, because it doesn't allow for the kinds of correlations these problems have between acts and the states of the world that are independent of one's acts. Jeffrey's evidential decision theory recommends not smoking in Fisher's case and cooperating with the other prisoner in the Prisoner's Dilemma, because the correlation with the big benefit is more important than the causation of the small one.

2.5.3 Causal and functional decision theories

Starting in the 1970s, many philosophers tried to develop versions of "causal decision theory" that appropriately represent the distinction between spurious correlations and ones that actually indicate the causal influence of one's actions. These theories recommend smoking in Fisher's case and defecting in the Prisoner's Dilemma, because they ignore the correlations between act and state and just recommend trying to cause the slight benefit that is still possible to cause. Representation theorems that develop this sort of theory are given by Brad Armendt (1986) and James Joyce (1999). Again, the point of the representation theorem is to show that anyone who makes normatively plausible decisions must implicitly contain some structure analogous to degree of belief that obeys the probability axioms. But even these causal decision theories have cases that are alleged to be counterexamples to them (Egan, 2007 and Ahmed, 2013).

However, in recent years, yet another theory has been developed, known as "functional decision theory" (Yudkowsky, 2010, Yudkowsky and Soares, 2018, and Soares and Levinstein, 2019), which recommends smoking in Fisher's case, but cooperating in the Prisoner's Dilemma. I argued in Easwaran (2021) that one of the best ways to understand this theory, and others like it, is by thinking about what the common cause is that explains the correlation between one's act and the state that brings about the good outcome. In Fisher's case, this common cause seems to be genetic, and thus outside one's control even in a very generalized sense, while in the Prisoner's Dilemma, the common cause between one's own behavior and one's co-conspirator's behavior was at least under one's control at some time in the past. Functional decision theory effectively sees the agent not as a physical being in the world, but rather as an abstract and timeless decision algorithm (which may be implemented by one's physical body, and perhaps by other physical beings as well). This perspective is perhaps more plausible for artificial intelligences than for humans, but many people feel that functional decision theory gives more intuitive recommendations than either evidential or causal decision theory. (Though Easwaran et al. (2026) argue that causal decision theory can actually mimic the preferences of functional decision theory, if we look at the more complex sorts of causation that exist in repeated interactions.)

However, at this point, functional decision theory has been given no representation theorem and it can be developed only if we assume that there is some prior concept of degree of belief that can be measured numerically. Thus, it may not be a theory of pragmatic decision making that can give a proper grounding for probabilistic degrees of belief.

Causal decision theory is sometimes said to have an advantage over evidential decision theory in that it respects dominance reasoning—at the point in the Prisoner's Dilemma that one is deciding whether to cooperate or to defect, we can treat the other player's act as a fixed external state, and regardless of which act they do, one's own act of defection yields a better outcome. However, this relies on Savage's conception of a "state" as something that is outside of the agent's control. Much of the debate between causal, evidential, and functional decision theory concerns precisely the question of what it is to "control" something.

The functional decision theorist, in particular, thinks of the agent as a multiply-realized decision algorithm. Although one *body* has no causal control over what the other *body* does, if the bodies are known (or even suspected) to operate according to some shared decision algorithm, then the outputs of this algorithm control *both* bodies. The algorithm leading to a choice of defection would not in this case be a dominant outcome, because the algorithm leading to a choice of cooperation would lead to a better outcome for both players.

While the evidential decision theorist doesn't have this particular conception of the agent, they still dispute the relevance of the Savage-style "states" that the causal decision theorist is implicitly relying on in the dominance argument. The proposition that the agent cooperates, and the proposition that they defect, are true in disjoint sets of worlds. There is no privileged identification of the worlds in one proposition with the worlds in the other proposition. The causal decision theorist's dominance argument privileges the identification of worlds in one of these propositions where the other player defects with worlds in the other proposition where the other player defects, but the evidential decision theorist could also point to an identification of worlds between the two propositions based on whether the other player does the *same* or *different* act as the agent. The former identification may preserve the causal powers of the agent, but the latter preserves the evidential probabilities. In the latter identification, neither act is dominant—cooperation does better in the states where the other player does the same, while defecting does better in the states where the other player does the opposite.

2.6 CONSTRUCTIVISM VS. REALISM

This brings us back to a central issue that can be raised independently of the distinctions between event and act and between causation and correlation. In the simple Dutch book argument and in Ramsey and Savage's representation theorems, it doesn't matter what *sort* of cognitive, behavioral, or biological mechanisms the agent uses to govern their preferences and behaviors—all that matters is that they *do* govern them in a way that avoids the kinds of violations that the arguments consider. Degrees of belief and utilities can then be constructed as a useful way of talking about certain patterns in these preferences, whether or not there is some explicit representation of them in the agent.

But many philosophers are specifically interested in the kinds of introspectible mental states that we humans do seem to have. Even though Ramsey noted that it is often hard to be sure which propositions one is more or less confident of, we do have some sense of this. And many of the different sorts of decision theory that I have mentioned, including the basic expected value that arises in Savage's theory, the risk-weighted expected value of Buchak, and the evidential, causal, and functional decision theories for dealing with act-state correlations, are much easier to understand if we start with already-understood concepts of utility and degree of belief, rather than constructing them as whatever it is that happens to make one's preferences work. I will consider this idea more closely in Section 3.6.1.

However, the perspective on utility and degree of belief given by the representation theorems is especially helpful when talking about doxastic agents that aren't like humans. And thinking about a variety of such agents and how they control their actions in consistent ways can help us imagine different pictures of how the human mind might work, since its fundamental workings are still not fully understood. We can imagine a corporation whose policies make it exceptionally clear what the degrees of belief and utilities are. Its charter might precisely spell out exactly how its employees should consider tradeoffs between profits now and profits in the future, and it could require its employees to maintain a database with an exhaustive listing of possible states of the world and outcomes of its decisions, and how much weight the corporation should give each of these possibilities in its actions. For this corporation, the degrees of belief derived from the representation theorem would correspond almost exactly to the numbers stored in the database, and the utilities derived from the representation theorem would correspond almost precisely to the tradeoffs spelled out in the charter.

But we could also imagine an organization that works somewhat differently. Consider a spy agency that behaves in many ways like the corporation we just considered—but imagine that there is a particular sensitive set of topics that its employees know to never incorporate into the database. As long as there is tight enough communication between the employees of this spy agency, they can keep their behavior coordinated in a way that still consistently furthers the goals of the agency. The representation theorems will still postulate degrees of belief and utilities for this agency, just as for the corporation. But these won't fully correspond to the information stored in the database—instead, there will be two types of representation of doxastic possibilities, one stored in the database, and the other just stored in the heads of the employees.

In a different direction of sophistication, we could imagine a simple animal that governs its actions purely through the instincts programmed into its biological neural structure, or the artificial neural net of a social media network's recommendation algorithm. These sorts of neural structures are famously difficult to interpret, and often don't have any separable sub-systems that correspond to the sorts of things that we think of as parts of their behavior. They may not have anything like the explicit representations of possibilities, propositions, preferences, utilities, or degrees of belief that some of the earlier systems do. But they do have some sort of complex control of their behaviors that will implicitly define preferences. And if these preferences satisfy the representation theorems (or even roughly do, as Neth (2024) suggests) then we can use this to get a grip on some pragmatic concepts of utility and degree of belief for these agents.

It's an open question in cognitive science, the extent to which the human mind is more like one or another of these systems or the extent to which it works in its own extremely different ways. But the example of the spy agency, and Ramsey's observations about the difficulty of introspecting some of one's deepest beliefs, suggest that there could be a lot more to our own minds than we think. While there might be some explicit representations of some degrees of belief and some utilities, there might be other aspects of our mind that play some of the same doxastic roles as these that we aren't aware of.

Economists have found it very productive to measure utilities for humans by means of "revealed preference". Although this idea was developed by Paul Samuelson (1938, 1948) initially as an *alternative* to utility theory, this was partly because he was reacting to versions of utility theory that hadn't adopted the representation theorems yet. The basic idea is that we can better estimate the value people put on things by observing their behavior than by asking them about their preferences verbally. Some of the most well-known studies investigate the "value of a statistical life", by seeing how the wages of similar low-skill jobs relate to the risk of mortality in those jobs. The idea is that since these are all similar jobs that don't require any specialized skills, the difference in wages that workers receive must mostly be compensating people for the risk of death. These studies often estimate the value of life in the range of several million dollars, which is quite a bit higher than the numbers that people give when explicitly asked about public health policies with known mortality impacts (Colmer, 2020). Some of these differences may naturally be explained because people have a better understanding of monetary amounts at the level of their wages, and risks they face on known job sites than they do of the monetary amounts discussed at the population level, and risks framed numerically in the context of a pandemic.

On the functionalist account, credence and utility just are whatever features of an agent happen to cooperate in the appropriate way to give rise to their actions. For some creatures, these will be real representations stored discretely within them, while for others, these will be abstract constructions out of the preferences that correspond only roughly to anything stored in a systematic way. The fact that the system as a whole has the behavior that it does means that there is something real that is useful to talk about as degree of belief or utility, though it might not be anything that one can identify in some way other than by looking at the behavior.

2.7 CONCLUSION

The distinctive feature of an agent is that they are a being that acts in the world in something like a goal-directed way. To the extent that they are effective at this, they can be understood as having a doxastic state that satisfies the probability axioms. If they are willing to buy and sell bets on propositions using some numerical value like money, and they avoid combinations of bets that guarantee that they end up worse off, then their degrees of belief can be read directly from the prices they are willing to buy or sell these bets at. If they have dispositions for accepting offers by someone who convinces them they can control the entire world, then

Ramsey has shown how to read off their utilities for worlds, and degrees of belief for propositions. If they have dispositions to prefer acts that yield various consequences in states independent of their agency, and those dispositions satisfy Savage's axioms, then Savage has shown how to read off their degrees of belief and utilities, and also that all of their preferences correspond to those given by expected value. Buchak, Jeffrey, and Joyce have given representation theorems that slightly modify these axioms to allow for risk aversion and for act-state correlations. Whether one of these sets of axioms gives the proper normative foundation for action or whether each of them is appropriate for a slightly different conception of an agent is still controversial.

This resulting picture of the doxastic state for an agent doesn't require any significant independent reality for the degrees of belief—they might exist prior to the agent's actions or they might just be a theorist's construction. Thus, basing a Bayesian epistemology on this pragmatic role for the doxastic state can allow us to make sense of doxastic states even for animals, artificial intelligences, and agencies that operate in a highly distributed way with no central representations that correspond directly to what we think of introspectively as our own beliefs.

All of these arguments show that probabilistic representability is essential for a certain kind of consistency in the agent's preferences among actions. An agent who violates them will do badly in a particular way—they will sometimes prefer a combination of acts that *they can tell* will guarantee that things end up worse off according to what they value. But satisfying these axioms is not a guarantee that the agent will be successful. No agent is in a position to tell with certainty what the result of their actions will be. Sometimes, the best act has a chance of turning out very well, but also a risk of turning out badly, and nothing can eliminate this risk. An agent will generally be more successful if it has high degrees of belief in true propositions and low degrees of belief in false propositions. But nothing we have said so far ensures that this will be the case, and the various forms of radical doubt going back to Pyrrho, Zhuangzi, Al Ghazali, Descartes, Hume, and others show that nothing can.

However, in Chapter 6, when I discuss the way that degrees of belief update, I will show that an agent that is internally consistent enough to count as Bayesian will tend to converge toward the truth. And in Chapter 8, I will further discuss the prospects for ensuring that one's degrees of belief in fact lead to successful action and are in fact accurate. But in Chapter 3, I will discuss the prospects for a different sort of doxastic subject, whose degrees of belief have a primary goal of aiming at the truth, rather than guiding any kind of action.

RESEARCH QUESTIONS

1. Is there a better normative characterization of belief or degree of belief in terms of the way it guides preference, action, or behavior than the ones given by Ramsey, Savage, Buchak, Joyce, and others?

2. Is there a better metaphysical characterization of acts as objects of preference, other than Savage's view of acts as arbitrary functions from mind-independent states to consequences or Jeffrey's view of acts as propositions just like any other?

3. Do Savage and Jeffrey's theories adequately deal with infinity?

4. Are there further modifications of Buchak's theory to deal better with risk-sensitivity or is risk-sensitivity irrational?

5. Are there further modifications of evidential, causal, or functional decision theory to adequately deal with act-state correlation?

6. Is there a good mathematical formalism for combining risk-sensitivity with act-state correlation?

CHAPTER 3

Belief aims at the truth

3.1 TRUTH, EVIDENCE, AND ACTION

In the previous chapter, we saw that by considering the role that degree of belief plays in explaining action, we can argue that rational degrees of degree of belief ought to be representable numerically in a way that satisfies the probability axioms. We don't need to assume there is a discrete feature of the agent that counts as their degrees of belief—as long as they are guiding their actions in a self-consistent way, we can represent them in terms of degree of belief and talk about their pragmatic behavior by talking about these degrees of belief. However, for a variety of reasons, some philosophers have thought that this is inadequate as an account of requirements of epistemic rationality.

Some philosophers, like Skyrms (1984) and Christensen (1996), have thought that for creatures with identifiable degrees of belief, although the practical irrationality of taking actions that in fact lead to bad consequences is not the same as doxastic irrationality, it can still be used as a *symptom* to diagnose doxastic irrationality. If doxastic states function as evaluations of the value of potential actions, then one might say that doxastic states that would lead to actions with *guaranteed* bad consequences amount to inconsistent evaluations of the same action. Because the actions are preferred, they are endorsed as good, but because the agent sees no possibility on which the consequence is good, these actions are also evaluated as bad. The fact that these doxastic states lead to two inconsistent evaluations of the same action serves as a diagnosis of a truly *epistemic* sort of badness. This is meant to be different from the *practical* badness of the fact that these doxastic states justify acts that, from the agent's own perspective, are guaranteed to lead to a bad consequence. Even Ramsey (1926, p. 182) says something similar, characterizing the

DOI: 10.4324/9781315626918-3

Dutch book argument as showing that the probability axioms amount to "laws of consistency" for credence, because a violation of them amounts to evaluation of the same set of actions in inconsistent ways that "depend on the precise form in which the options were offered".

I think that there is no need for this reinterpretation, if the doxastic state really serves a primarily pragmatic function. If all there is to "degrees of belief" is some sort of abstract characterization of however it is that the agent guides their actions, and if this guide to actions is bad on its own terms, then these degrees of belief are irrational. This is totally sufficient for agents whose normative requirements are all fundamentally pragmatic. I think it's very plausible to include humans in this category.

But some philosophers think this is insufficient. They want to sharply distinguish practical rationality from epistemic rationality. They think that, to properly count as a doxastic state, it must be subject to some epistemic requirements. Skyrms, Christensen, and Ramsey's reinterpretation of the pragmatic arguments as demonstrating an inconsistency are an attempt to satisfy this concern. But I think that a proper account of this sort would have to identify the core function of the doxastic state as something non-pragmatic.

The classical pragmatic arguments for probabilism described in the previous chapter arise originally in the work of Ramsey (particularly his (1926)) and de Finetti ((1931/1989) and (1937/1964)), who follow this track because of several philosophical commitments. One is a commitment to a kind of pragmatism, that thinks of agents as inherently playing a role of *acting* or *doing* things in the world, with all other roles (including doxastic) being understood in terms of this. Additionally, Ramsey and de Finetti were working at the heyday of positivism, operationalism, and behaviorism, a set of related movements in philosophy and science that required a concept to be observable in order to be meaningful. For mental states in particular, this often meant that the state itself was cashed out in terms of the behaviors or acts that it would lead to.

Contemporary philosophers often don't share either of these commitments. For instance, Eriksson and Hájek (2007) reject the very idea that action could be central to degree of belief. On p. 194 they suggest the conceivability of a kind of Zen or Stoic practice that eliminates all desire (and thus all value for action) while maintaining doxastic states. Whether or not this is actually possible for us humans, someone motivated by a broader interest in epistemology for *all* kinds of beings (such as scientific communities or artificial intelligences or aliens) might be interested in an account of doxastic states that doesn't conceptually tie them to action. Similarly, since the cognitive revolution of the 1950s, most psychologists and philosophers have rejected behaviorism and are thus far more willing to postulate mental states not essentially connected to any observable behavior. If these mental states have their own existence independent of the actions that they help guide, we can then consider what aims they might have other than practical success. (In Section 3.6.1, I will show how these states might still be useful for guiding action, even if that is not their fundamental role.)

Thinking of doxastic states as representational states with a mind-to-world direction of fit has led to several ideas about how to derive norms for these states. In recent years, many philosophers working on these ideas have been inspired by the debate between William Clifford (1877) and William James (1896). Even though both of these philosophers shared some versions of the pragmatist commitment, that the point of belief is to guide action, they still suggested forms of epistemic evaluation that many anti-pragmatist philosophers consider acceptable.

Clifford begins his argument by saying that since beliefs have practical consequences in the form of action, their evaluation is actually a moral affair. However, he considers a case in which a person risks the lives of others on insufficient evidence, and turns out lucky that no one is hurt. He says,

> The man would not have been innocent, he would only have been not found out. The question of right or wrong has to do with the origin of his belief, not the matter of it; not what it was, but how he got it; not whether it turned out to be true or false, but whether he had a right to believe on such evidence as was before him.

He eventually summarizes his point by saying "it is wrong always, everywhere, and for anyone, to believe anything upon insufficient evidence."

James, however, disagrees with Clifford's assessment. He claims that there are certain truths that Clifford's rule would cut us off from permanently. He notes that in science, it is a commonplace that a theorist must usually believe their theory before they will have the proper motivation to persist through the difficulties of experimentation to eventually provide the evidence that will some day support the theory. And more importantly for him, he claims that when it comes to social beliefs (that a stranger is friendly) or religious beliefs (that the universe has some spiritual nature), it may be literally impossible to get the evidence if one doesn't already have the belief that makes one open to the evidence.

Since these beliefs can be quite momentous, he says that an adequate normative theory for belief must not close off the permissibility of such beliefs by requiring evidence to come before the belief.

> Our great difference from the scholastic lies in the way we face. The strength of his system lies in the principles, the origin, the *terminus a quo* of his thought; for us the strength is in the outcome, the upshot, the *terminus ad quem.* Not where it comes from but what it leads to is to decide.

Although James, as a pragmatist, does think that beliefs must eventually lead to action, in this paper he focuses on the achievement of truth itself.

> We must know the truth; and we must avoid error,— these are our first and great commandments as would-be knowers; but they are not two ways of stating an identical commandment, they are two separable laws.

He characterizes Clifford as being too concerned with avoiding error to allow himself to know the truth, and says instead that different people might have a "passional nature" that puts different weight on the importance of knowing the truth and the importance of avoiding error.

This chapter investigates various ways of spelling out these different roles for belief or degree of belief. Rather than thinking of them in terms of their pragmatic role of guiding action, we can think of them in terms of their alethic role of aiming to be close to the truth (as James focused on), or their role of following the evidence (as Clifford did). Section 3.2 considers the alethic role for belief and Section 3.3 considers the alethic role for degree of belief. Section 3.4 considers challenges to these ways of grounding probabilism. Section 3.5 considers the evidential role for the doxastic state and then explores the prospects for understanding belief or degree of belief in terms of an evidential role. Finally, Sections 3.6.1 and 3.6.2 consider how these different roles for degree of belief might interact with the pragmatic role.

3.2 THE ALETHIC ROLE FOR FULL BELIEF

3.2.1 Hempel's account, with logical probability

The simplest way to formalize James's account is to consider a subject with an attitude of belief and a set of propositions that they can either bear this attitude to or not. We say that the overall "epistemic value" of the subject's doxastic state is the number of true propositions they bear this attitude to minus the number of false propositions they bear it to. Carl Hempel investigates a version of this account toward the end of his long paper on scientific explanation (pp. 153-155 of his (1962)). Hempel assumes that there is some "inductive probability" available to guide the subject in forming their beliefs. This inductive probability need not be any feature of the subject themself—he probably thinks of it more along the lines of the logical concept of probability that Ramsey attacked. But Hempel does assume that this logical probability makes it rational to have particular beliefs.

In particular, he uses the same expected value formula discussed in Section 2.3.5. Since the epistemic value of the total doxastic state is just the sum of the value of the attitude for each proposition (either 1 for believing a true proposition, -1 for believing a false proposition, or 0 for not believing a proposition), the linearity of expectations means that a subject maximizes their overall expected epistemic value if and only if they have the attitude to each proposition that maximizes expected value for that proposition. If P is the inductive probability function, then the expected value of believing a proposition p is $1 \cdot P(p) + (-1) \cdot P(\bar{p}) = P(p) - (1 - P(p)) = 2P(p) - 1$, while the expected value of not believing it is 0. Thus, he argues that, with this way of aiming at the truth, a rational subject will believe a proposition if and only if its inductive probability is greater than 1/2. The result is the same as long as the value of a true belief is just as much higher than the value of no belief as the value of a false belief is below it.

3.2.2 The Lockean thesis for belief with degree of belief

Dorst (2019) gives a slightly more sophisticated version of this picture in which the epistemic benefit for believing a truth can differ from the epistemic cost of believing a falsehood, perhaps based on the "passional nature" of the individual, as suggested by William James. We can call the value of believing a truth "R" for the value of being "right" and the value of believing a falsehood "$-W$" for the value of being "wrong". From a similar calculation to Hempel's, Dorst shows that one should believe a proposition if its probability is above $W/(R+W)$, believe its negation if the probability is below $R/(R+W)$, and neither if the probability is in between. (He argues that W should in fact be greater than R so that belief is consistent. Shear and Fitelson (2019) argue further that the ratio W/R should be greater than the Golden Ratio $(1+\sqrt{5})/2$ to achieve other desirable properties of belief!)

Both of these accounts show how the search for the truth can constrain belief if there is a probability function already available, and if maximizing the expected value of belief with respect to this probability function is essential for doxastic rationality. Hempel appeals to a logical probability function, whose existence is now quite controversial. But if both belief and degree of belief are real parts of a subject's doxastic state, and if degree of belief plays the role of guiding their actions (as discussed in the previous chapter), and if belief is like an action that degree of belief can control, then this results in an argument for the Lockean thesis discussed in Section 1.5.1.2. This requires a pragmatic role for degree of belief and an alethic role for belief, and it requires degree of belief to be prior to belief.

3.2.3 Belief without degree of belief

However, we can also imagine doxastic subjects that have access to a set of propositions, and an attitude of belief that they can hold toward them *without* having degrees of belief. This could be the situation of certain kinds of artificial intelligences that work with propositions written in a database or groups that work with a kind of constitution or credo spelled out in words, but without qualifiers for degree.

I develop this idea in Easwaran (2015), relying only on the alethic values R and W of beliefs being right or wrong, without any input from credence. For a subject with this kind of doxastic state, it's clear that the best overall pattern of beliefs to have would be one that believes all true propositions and no false ones. However, no finite subject, even if it is constituted by a powerful computer or an enormous group of people working together, is able to ensure that they do this, if the set of propositions they are considering is moderately large. There will generally be some propositions for which believing it runs the risk of being wrong, and withholding belief runs the risk of not being right. Without a full decision theory for belief, we can't say a lot about what *specific* set of beliefs a subject should have on the basis of their imperfect information and their passional nature. But it turns out that we can say some things based on a principle of "dominance" or "admissibility"—if one option is guaranteed to do at least as well as another, and sometimes better, then the first is said to dominate the second, and the second is said to be inadmissible. The idea is that one shouldn't have a belief state that is inadmissible, regardless of how one deals with cases that are uncertain.

For instance, if we consider a proposition p and its complement $\bar{p}$, then we can see that believing both gives an epistemic value of $R-W$ for sure, while believing neither gives a score of 0 for sure. By dominance, we can see that if $R < W$, then it's worse to believe both p and $\bar{p}$ than to believe neither. This doesn't mean that believing neither is the best option overall—in fact, if p is true, the best option is believing just p, and otherwise the best option is believing just $\bar{p}$. But if $R < W$, then we can say that one definitely shouldn't believe both. There is one option of believing neither, which is clearly better regardless of what the truth is, even if we can't identify the best possible option.

This criterion of admissibility leads to several other constraints on belief as well. If the subject takes p to entail q, so that there are no possibilities on which p is true without q being true, then the subject's doxastic state would be doing a bad job if it believed p without q. That's because believing q without p would dominate this belief set. In a possibility where both p and q are true or both p and q are false, these two belief sets get the same alethic value, while in a possibility where p is true and q is false, the state of believing p but not q does better than the reverse. One can't say whether it's best to believe both, or neither, or q but not p, but one can rule out one set of doxastic attitudes.

By eliminating inadmissible sets of attitudes, we can get some constraints on what a subject should believe, but these are "wide-scope" in the sense of Broome (2007)—we show that certain combinations of attitudes are bad, but don't say which particular attitude should be changed if the subject has some that conflict. If p entails q, and the subject finds herself believing p but not q, we don't say that the subject *should* believe q or *should* withhold belief from p—we just say that the subject recognizes a particular option (namely believing q but not p) to be better than their own, and thus say that it is irrational from their own point of view to believe p but not q.

In fact, in Easwaran (2015), I conjectured that this might be able to ground a concept of degree of belief for such a creature, and Rothschild (2021) shows that this is correct. Imagine that the absolute value of R and W is not the same for every proposition, but their *ratio* is constant, though some propositions have more "weight", and thus give both more benefit if believed while true and more cost if believed while false. If there is a probability function such that the subject satisfies the Lockean thesis with respect to that probability function, then that set of beliefs maximizes expected alethic value, and thus no other doxastic state could dominate theirs. But if there is no probability function such that the subject's beliefs satisfy the Lockean thesis with respect to it, then there is a different set of beliefs that, on some assignment of weights to the propositions, dominates their beliefs. If the subject's beliefs are doing a good job of satisfying their alethic role, then there shouldn't be any dominating set of beliefs. So there should be such a probability function. We can then take this probability function to be a representation of the subject's "degrees of belief", even though fundamentally, their doxastic state just consists of beliefs. This is akin to the way degrees of belief are constructed from an agent's preferences in the representation theorems of the previous chapter.

3.3 ACCURACY AND DEGREE OF BELIEF

This discussion of belief aiming at the truth is simpler than the situation for degrees of belief, but much of it was inspired by much more precise dominance arguments given about degrees of belief, which I will now describe. The philosophical evaluation of the alethic role of degree of belief dates back to Joyce (1998), though the mathematical arguments he uses go back to de Finetti (1974) and Lindley (1982). In the case of belief, for each proposition there is just a possibility of believing or not, and a belief can only be right or wrong. The only interesting variation for the evaluation is the ratio of the value for rightness and the penalty for wrongness. But with degree of belief, there are many levels in between the extremes, and these intermediate degrees of belief can't be said to be "right" or "wrong".

There has been less work justifying the idea that alethic value can be measured on a numerical scale than there has been for practical value. But if we imagine a subject with no practical desires to act, but a concern for alethic value, then it is plausible that some of the arguments that establish the numerical scale for utility might help justify a numerical scale for alethic value. (It's not obvious that this will work, since many of the constructions of the numerical scale for utility require degrees of belief that are separate from the acts, and in this case, the degrees of belief roughly *are* the acts.) I will arbitrarily assume that the subject's doxastic state can be represented by a single degree of belief in each proposition. I will argue that under certain conditions, this degree of belief can be represented numerically, but I don't take any numerical features to be essential to the degree of belief itself.

Many authors here make the assumption that the overall alethic value of a doxastic state is just the sum of the alethic values for the agent's attitude to each proposition individually, and that an attitude to one proposition is evaluated in the same way that the same attitude to a different proposition is. I will do the same for most of this section.

3.3.1 Numerical degrees of belief

In the pragmatic case, we argued that we could represent the degrees of belief in each proposition by a single real number because we could use the prices of various bets that the agent was willing to accept. But for a subject whose degrees of belief are just evaluated alethically, there are no such bets. However, we have assumed that each proposition makes the same type of contribution to the overall evaluation of the doxastic state, and that a given degree of belief is evaluated the same way for each proposition. So we can identify degrees of belief by the contributions they make to the alethic evaluation of the doxastic state.

Because this evaluation is alethic, there is one contribution that a given attitude makes when the proposition the subject has it to is true, and another contribution that the attitude makes when the proposition the subject has it to is false. So any degree of belief can be identified by these two numbers. If there are two distinct degrees of belief that have the same two numbers (that is, they make the same contribution to the doxastic state when the relevant proposition is true, and they make the same contribution to the doxastic state when the relevant proposition is false), then there is no normative distinction between these degrees of belief. Given the functional characterization of degree of belief we are working with (as discussed in Section 1.2.3) these would effectively be the same degree of belief. Thus, the degrees of belief can be identified by at most two numbers.

But in fact, we can do better. One of the numbers says how good or bad the degree of belief is for a proposition that is true, and the other says how good or bad the degree of belief is for a proposition that is false. If x and y are two different degrees of belief, but x is better than y *both* when the proposition is true *and* when the proposition is false, then y just isn't a degree of belief that any subject should ever have—x would always be better. Similarly, if x is *worse* than y both for true and false propositions, then x isn't a degree of belief that any subject should ever have. In fact, even if x and y are equally good when the proposition is true, then whichever one is worse when a proposition is false would never be a good one to have. Thus, for any two degrees of belief that can both be worth having, one of them must be better when the proposition is true, and the other must be better when the proposition is false. The set of relevant degrees of belief is in fact one-dimensional, just as in the pragmatic case.

It is standard to arbitrarily represent the degree of belief that is best to have when a proposition is true with the number 1, and the degree of belief that is best to have when a proposition is false with the number 0, and the other degrees of belief with numbers in between. There is still nothing as clear-cut as the prices of bets to decide *which* number is used to represent which degree of belief, but I will discuss that issue more explicitly in Section 4.1.3.3.

3.3.2 Inaccuracy functions and propriety

Once we have identified each degree of belief with a real number between 0 and 1, we need a way to represent the value of having each degree of belief when the proposition is true and when the proposition is false. We could choose to represent these alethic values in such a way that positive numbers are good and negative numbers are bad, but it turns out to be a bit more convenient to do the reverse. Thus, we think of alethic value as measuring the *inaccuracy* of a degree of belief. I will use $I(x, 0)$ to represent the function that says how bad it is to have degree of belief x in a proposition that is false, and $I(x, 1)$ to represent the function that says how bad it is to have degree of belief x in a proposition that is true. It is standard to set the scale such that $I(0, 0) = 0$ and $I(1, 1) = 0$. These inaccuracy functions are often called "loss functions" or "scoring rules" when they are studied in the context of statistics or machine learning.

The inaccuracy function that is discussed most often in the philosophical literature comes from Brier (1950) and is often called the "Brier score" or "quadratic loss". This inaccuracy function is given by $I_{Brier}(x, 0) = x^2$ and $I_{Brier}(x, 1) = (1 - x)^2$. It was initially proposed as a way of incentivizing meteorologists to do the best job they could at predicting the probability of rain, but in this context it depended on a purely pragmatic understanding of the role of degree of belief. The idea was that meteorologists should be penalized for reporting high probability of rain if it doesn't rain and penalized for reporting low probability of rain if it does rain, so that they would have motivation to use all the information they could to try to figure out what would happen. But if the penalties were chosen incorrectly, the meteorologist might not report honestly.

If we assume that a meteorologist has some private degree of belief x that it will rain, but they report degree of belief y, then we can calculate their expected penalty as

$$xI(y, 1) + (1 - x)I(y, 0).$$

This is because $I(y, 1)$ is the penalty they receive for reporting y if it turns out to rain, and x is their degree of belief that it will rain, and similarly for $I(y, 0)$ and $1-x$. So Brier wanted to find a pair of functions $I(y, 0), I(y, 1)$ such that $xI(y, 1)+(1-x)I(y, 0) \geq xI(x, 1)+(1-x)I(x, 0)$, with equality only when $y = x$. Reporting their own degrees of belief would be the way that a meteorologist would minimize their expected penalty. This property of an inaccuracy function is called "propriety", and such a function is called a "proper scoring rule". It's clear that propriety is essential for this *practical* purpose, but I will discuss its motivation for a measure of properly *alethic* value in Section 3.4.3.

Under the Brier score, the expected inaccuracy for reporting y is

$$x(1 - y)^2 + (1 - x)y^2 = x - 2xy + xy^2 + y^2 - xy^2 = x - 2xy + y^2.$$

If the meteorologist were to report x in particular, we can see that their expected inaccuracy would be $x - x^2$. The difference between these two is

$$(x - 2xy + y^2) - (x - x^2) = x^2 - 2xy + y^2 = (x - y)^2.$$

Thus, we can see that whenever $y \neq x$, the expected inaccuracy of reporting y must be strictly greater than the expected inaccuracy of reporting x.

But this is not the only proper scoring rule. The other proper scoring rule that is most often discussed is the logarithmic rule, with $I_{Log}(x, 0) = -\log(1 - x)$ and $I_{Log}(x, 1) = -\log x$. (It doesn't matter whether the logarithm is calculated with base 10, or 2, or e, or any other number—the conversion from one to another corresponds to multiplying by a constant.) For this scoring rule, the expected inaccuracy of reporting y is

$$-x \log y - (1 - x) \log(1 - y).$$

To show that this is strictly proper, we can use some calculus and see that the derivative of this expected inaccuracy with respect to y is $-\frac{x}{y} + \frac{1-x}{1-y}$. When $y < x$, the first term is greater than 1 while the second term is less

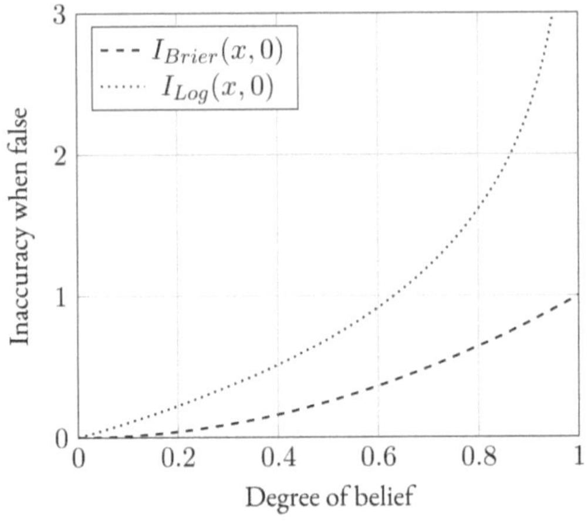

Figure 3.1 Comparison of $I_{Brier}(x, 0) = x^2$ and $I_{Log}(x, 0) = -\log(1 - x)$.

than 1, so the derivative is negative, while when $y > x$, the first term is less than 1 while the second term is greater than 1, so the derivative is positive. Thus, again, expected inaccuracy is minimized when reporting $y = x$.

There are many other proper scoring rules as well, though most of the others are less discussed in the philosophical literature. Predd et al. (2009) give a characterization of the broader class of them in terms of "Bregman divergences". The main difference between different proper scoring rules is that they make different degrees of belief contribute more or less to the subject's overall inaccuracy—no degree of belief can ever have more than four times the inaccuracy of 0.5 on I_{Brier}, while on I_{Log}, inaccuracy can go to infinity for degrees of belief that are close to the wrong end of the scale, as shown in Figure 3.1.

If we follow William James, we might say that an agent whose doxastic state plays an alethic role just has some particular "passional nature" that determines how bad it is for their degrees of belief to be inaccurate in one way or another, and that this determines which proper scoring rule is relevant for evaluating their degrees of belief. But for artificial intelligences, the choice might be up to the designer, as I will discuss further in Section 3.3.3. Some arguments for probabilism depend on the particular measure of inaccuracy that plays the relevant normative role, but others are more general.

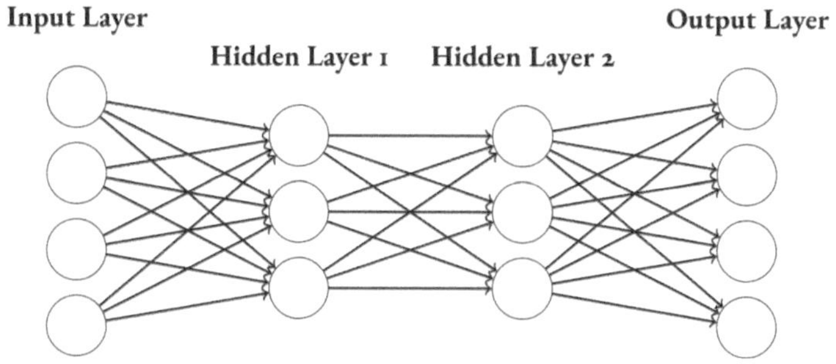

Figure 3.2 A simple neural net, with four input neurons, three neurons in each of two hidden layers, and four output neurons.

3.3.3 Neural nets as purely alethic subjects

Before giving the argument for probabilism, it is worth considering concrete examples of subjects that might be said to have degrees of belief subject to alethic evaluation according to a proper scoring rule. One of the best examples is an artificial neural net trained on a classification task.

The full details of a neural net aren't essential for our purposes. But the relevant features for our purposes are these. The network consists of a set of neurons, organized into a sequence of layers, with the first layer called the "input neurons" and the last layer called the "output neurons", while intermediate layers are called "hidden layers". On a given run of the neural network, the input neurons are initialized with some real number values, and then the neurons in each layer sequentially calculate values as a function of the values of the neurons in the previous layer, and the values of the output neurons are then read out for some purpose. In many cases of interest, the value of each neuron is calculated by the application of the same "activation function" to an input determined by a weighted sum of the values of the neurons in the previous layer together with a bias unique to each neuron. The weights with which each neuron in one layer feeds as an input to each neuron in the next layer, and the biases of each individual neuron, are collectively known as the "parameters" of the network. This structure is illustrated in Figure 3.2.

In the example described by LeCun et al. (1989), there are 256 input neurons, each initialized with a number from -1 to 1 representing the grayscale shading of a pixel in a 16×16 scan of a handwritten digit. There are three hidden layers, each with several hundred neurons in it, with many of the neurons fixed to apply the same weights to different sets of neurons in the previous layer. There are 9760 total parameters in the network. The output layer consists of ten neurons, and the value of each at the end of the run will eventually be interpreted as the system's "degree of belief" that the input was a depiction of one of the ten digits: 0, 1, 2, 3, 4, 5, 6, 7, 8, 9.

The process of setting the final parameters in the network consists of several steps. First, they gathered scans of 7291 different handwritten digits whose classification was known to use as "training data". Then they started with a version of the network with randomized parameters. Then they repeatedly tested it on a small random subset of the training data, evaluated the Brier score of the resulting output neurons, and slightly modified the parameters to improve this Brier score. (The method for deciding how to modify the parameters is by calculating the derivative of the Brier score of the output with respect to each parameter. It turns out that the structure of network gives rise to a relatively simple algorithm for calculating these derivatives, known as "backpropagation of errors", as described by Rumelhart et al. (1986).) After approximately 23 passes of the network over all 7291 images in the training data, the average Brier score on these particular inputs was about 0.0025, meaning that it generally had degree of belief greater than .95 in the correct digit, and less than 0.05 in all of the other digits. Other 2007 images had been gathered at the same time as the training data, but kept out of the process—when the final network was tested on them, the average Brier score was slightly higher, but still quite low (about 0.018, indicating degree of belief usually above .9 in the correct digit).

Because of the initial randomization of the parameters of the network, and the randomization of the order of presentation of the training data as the parameters are updated, the final parameters of the network are not determined by this specification of the process. Rather, if someone else were to carry out the same process as described, they would end up with a different set of parameters, though it would likely also end up with relatively low Brier score. The structure of the training process makes the final parameters depend on their role in correctly classifying handwritten digits. Because of the role of the Brier score in this process, I claim this makes the features of the network normatively dependent on this particular measure of the alethic role. Thus, I claim that it is appropriate to think of these output neurons in the final trained network as representing degrees of belief that play the alethic role. On a given run of the neural net, the activations of the output neurons can be thought of as the system's "degrees of belief" that the input is each of the relevant digits.

LeCun et al. (2015) showed, a few decades later, that with more powerful modern computers, neural nets could be trained for much more general classification tasks, including classification of photographs as representing objects from one of thousands of different classes, and identifying appropriate words to use in the translation of a text from English to French. Because these tasks are so much harder, and some inputs are truly ambiguous, these networks more often had intermediate degrees of belief. Wang et al. (2022) and Terven et al. (2025) give surveys of how scoring rules other than the Brier score have been used in some of these applications. Vaswani et al. (2017) showed how, in particular, neural nets of these sorts could be used to create a system, called a "transformer", that has reasonable degrees of belief about which word comes next in texts drawn from the internet. Starting in late 2022, these transformers were incorporated into chatbots, like ChatGPT, Claude, and Gemini, that got further "fine-tuning", in ways that made them helpful to many people for a variety of purposes.

I don't specifically claim that the text outputs of these chatbots are appropriately interpreted as beliefs or other attitudes of the relevant systems, particularly if we are focusing on the alethic role. The process of fine-tuning the chatbots to provide helpful answers to queries, rather than just predicting which word comes next, changes the normativity they operate under. Many modern chatbots also undergo a process of reinforcement learning, in which their text outputs are used to generate solutions to problems with verifiable answers, and text that led to correct answers is reinforced while text that led to incorrect answers is downweighted. This may well make the outputs of the system (i.e., the text it produces) subject to a kind of pragmatic normativity rather than alethic. But it probably is appropriate to interpret the output of an image classifier or even the output of a transformer (interpreted as a prediction of what word comes next) as degrees of belief of the relevant system, subject to alethic evaluation.

I don't claim that the activations of neurons in the hidden layers of these neural nets are appropriately interpreted as degrees of belief of the system. The training process makes it very reasonable to interpret the output neurons as having propositional content—in the digit recognizer the training process aims to bring the output of the system closer to the truth value of the proposition "the current digit is x". But there is no direct control on the neurons in the hidden layers, and certainly no easy way to attribute content to them.

Goldstein and Levinstein (2024) argue that, despite the caveats I have given, it may well be possible to interpret some of the activations and weights on neurons in hidden layers of modern chatbots as beliefs or degrees of belief of the system. I agree with their claim that arguments that this *can't* be the case are mistaken. But I think that much more would need to be done to establish the propositional content of these attitudes, let alone establish that they really do have doxastic normative structure. And more relevantly for this chapter, it's not at all clear which of these attitudes might specifically play an alethic role, rather than a pragmatic one.

For doxastic subjects that aren't neural nets, I think parallel considerations apply. To the extent that animals, humans, and groups (and aliens and angels, if there are any) learn to form their degrees of belief purely through a process of training based on known data, subject to a given scoring rule, it is likely that they are alethic subjects. But most of them probably learn to form their degrees of belief in part by using these degrees of belief to govern their behavior. Thus, I suspect that in fact, most doxastic subjects are best interpreted as agents whose degrees of belief play a pragmatic role. But the rest of this section will continue to focus on the doxastic states for alethic subjects, and in particular show why their degrees of belief are subject to probabilism just as those of pragmatic agents are.

3.3.4 Probabilism on the Brier score

I will start by illustrating the argument for probabilism using the Brier score. This version of the argument happens to be one we can think of using Euclidean geometry. If there are n propositions that the agent has credences in, then we can think of their credences in these propositions as the coordinates of a point inside the n-dimensional cube. (The case where there are infinitely many propositions will be discussed briefly in Section 3.3.6.) Some of the vertices of this cube correspond to possible combinations of truth and falsity for all the propositions. (Not all of the vertices of the cube do, because there are logical relationships among some of the propositions, so some combinations of truth and falsity are impossible.) The Brier score of the subject's degrees of belief is the sum of the squares of the differences between those degrees of belief and the actual truth values of those propositions. Thanks to the Pythagorean

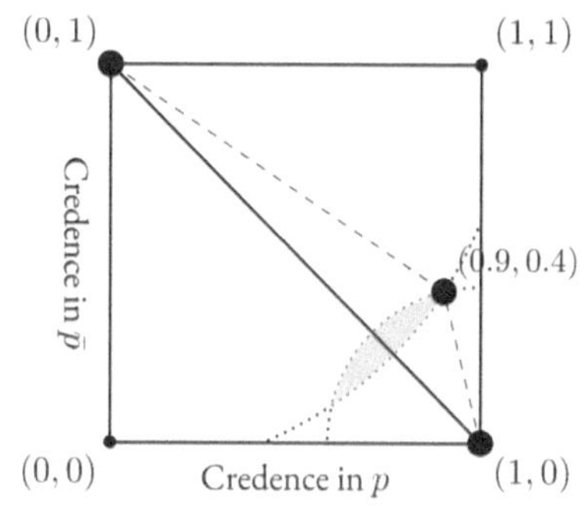

Figure 3.3 Diagram showing all possible doxastic states consisting of pairs of credences in p and $\bar{p}$. Dotted circles indicate doxastic states that are equally accurate to $(0.9, 0.4)$ when p is true (lower right) or when p is false (upper left). Shaded areas indicate credences that are strictly more accurate than it in both cases.

theorem, this sum of the squares of the individual differences is the square of the direct distance between that point inside the cube and the relevant vertex.

It's of course difficult to visualize an n-dimensional cube, but fortunately, the full n-dimensional cube isn't needed for any step in the argument. Figure 3.3 shows a cross-section through this cube that just visualizes the subject's possible degrees of belief in some proposition p and its complement $\bar{p}$. Consider what happens if the subject has degrees of belief .9 in p and .4 in $\bar{p}$. This subject's degrees of belief violate additivity (because these two propositions are disjoint, and the agent's credences in them add up to 1.3, rather than to 1, which should be their credence in the union). There are only two possible sets of truth values for these two propositions—either p is true and $\bar{p}$ is false or vice versa. So we can consider the circles around the corresponding vertices of the cube, which represent the other sets of degrees of belief that would be equally accurate in each of these possibilities. The lens-shaped region where these two circles intersect consists of the sets of degrees of belief that are more accurate than the subject's actual degrees of belief in both of these possibilities.

Thus, the subject is in a position to realize that there are alternative degrees of belief that are available, that would do better than theirs regardless of what the truth is. These degrees of belief are doing a bad job of satisfying their role of aiming at the truth, assuming that they are being evaluated on the Brier score. By looking at the geometry of the situation, we can in fact see that something similar will occur whenever the subject's degrees of belief in p and $\bar{p}$ fall anywhere off of the diagonal line between the two possibilities. But if the degrees of belief are precisely on that diagonal line, then the two circles will be tangent at the point, rather than having a lens-shaped intersection—the only degrees of belief that are closer to the truth in one possibility will be farther from the truth in the other.

This two-dimensional diagram only illustrates the argument for one particular type of additivity, between a proposition and its negation. But a very similar three-dimensional diagram (with spheres in place of circles) illustrates the argument for additivity generally. Non-negativity is automatic, because 0 is the lowest degree of belief on the representation we are using. And normalization is straightforward as well—is the subject has no doxastic possibilities where the proposition is false, then in every possibility, the best degree of belief to have is 1, and so having any other degree of belief in the proposition would be worse.

If one doxastic state is guaranteed to be more inaccurate than another, then we say that the one that is guaranteed to be more inaccurate is "inadmissible", because it is "dominated" by the other set. This argument shows that if inaccuracy is measured by the Brier score, then degrees of belief that violate the probability axioms are inadmissible.

Leitgeb and Pettigrew (2010a) actively use the geometric analogy to argue that the Brier score is the correct measure of inaccuracy for degrees of belief playing an alethic role. But most other philosophers using the accuracy framework have acknowledged that different subjects might have degrees of belief whose accuracy is measured by different functions and have tried to make the argument more generally.

3.3.5 Joyce's argument

Joyce (1998) does not assume that the Brier score is the one correct way to evaluate degrees of belief alethically. Instead, he gives an axiomatic characterization of how he argues an alethic measure of degrees of belief should behave, and then shows that a version of the same result holds regardless of the particular inaccuracy measure that is used. For any inaccuracy measure and any degrees of belief that fail to satisfy the probability axioms, there is another set of degrees of belief that has lower inaccuracy than those ones at every possibility. Violations of probabilism mean that a subject's degrees of belief aren't doing a good job alethically.

Joyce does *not* make the assumption I began with, that the inaccuracy of the overall doxastic state is the sum of the inaccuracies of each individual credence. However, he does assume that degrees of belief can be represented with real numbers and several of his axioms presuppose that this has been done in some particularly meaningful way. He doesn't make any explicit use of Euclidean distance, as we did when using the Brier score. But he does make use of some aspects of the geometric structure of n-dimensional space.

His first axiom, Structure, requires that the inaccuracy of a doxastic state is a continuous function of the degrees of belief in each proposition at each world. His second axiom, Extensionality, requires that this function not depend on the identities of the propositions—just the combinations of truth-values and degrees of belief. His third axiom, Dominance, states that the inaccuracy of a doxastic state at a world decreases if the degrees of belief of all but one proposition stay the same, while the degree of belief in that proposition gets closer to the truth value of that proposition at that world. In the context of an assumption that overall inaccuracy is the sum of inaccuracies for the degree of belief in each proposition, and the numerical representation of degrees of belief we chose that identifies higher numbers with degrees of belief that are better for true propositions, continuity is the only additional content of these first axioms. These axioms don't depend on the geometry of the degrees of belief, just the topological structure given by their ordering.

Joyce adds three more axioms. These ones do depend substantively on the numerical values of degrees of belief and they ensure that the measure of inaccuracy behaves somewhat like Euclidean distance. These seem much more problematic to me, since so little has been said about what justifies representing particular degrees of belief with particular real numbers. The axiom of Normality states that in addition to not depending on the identities of propositions, but only the degrees of belief and truth values, the inaccuracy of one doxastic state in one world is the same as the inaccuracy of a different doxastic state in another, if the degree of belief in any proposition whose truth value is different between the two worlds is "reversed". In the context of inaccuracy functions for individual propositions, this would mean that $I(x, 0) = I(1 - x, 1)$.

The last two axioms, Weak Convexity and Symmetry, both involve "mixtures" of two doxastic states. If one doxastic state has degree of belief x in p, and another has degree of belief y in p, then for λ between 0 and 1, the λ-mixture of the states has degree of belief $\lambda x + (1 - \lambda)y$ in p. Weak Convexity says that if two different doxastic states have equal inaccuracy in a particular world, then their 1/2 mixture has strictly less inaccuracy in this world. Symmetry states that if two different doxastic states have equal inaccuracy in a particular world, then their λ mixture and their $1 - \lambda$ mixture also have equal inaccuracy in this world (and it's not too hard to prove, given all the axioms, that the inaccuracy of these mixtures must also be strictly less than the inaccuracy of the initial doxastic states, with the 1/2 mixture being lowest).

Joyce then shows that a version of the same argument for the Brier score works. We can draw "spheres" in n-dimensional space around each possibility that indicate a set of doxastic states that all have equal inaccuracy at that possibility. If a doxastic state violates the probability axioms, then the spheres centered at each possibility that indicate the other doxastic states that are equally inaccurate as the original one will all have non-trivial intersection, and any doxastic state inside that intersection will be strictly closer to the truth at every possibility. But if a doxastic state satisfies the probability axioms, then these spheres all intersect only at that one doxastic state—any alternative doxastic state that is closer to the truth at one possibility must be farther at another. If there is an inaccuracy measure of this sort such that the role of a person's doxastic state is to be close to the truth by this measure, then degrees of belief that violate the probability axioms do a bad job at this role by the subject's own estimation, even without knowing which possibility is actual. Violations of the axioms of probability are inadmissible.

Joyce's argument identifies a set of features of the Brier score that suffice to ensure that probabilism is essential for playing the alethic role well. However, it's unclear how many measures of inaccuracy that are significantly different from the Brier score share these features. (Joyce shows that if, for each proposition p, there is a positive constant k_p such that the inaccuracy of having degree of belief x in p is given by $I_p(x, 0) = k_p x^2$ and $I_p(x, 1) = k_p(1 - x)^2$, and if the inaccuracy of the doxastic state as a whole is the sum of these inaccuracies, then the resulting measure satisfies his axioms. I am not aware of any proof that these are the only measures of inaccuracy that satisfy these six axioms, or if anyone has exhibited any others.) And Joyce's last three assumptions depend on a lot of detailed features of the numerical representation of degree of belief in ways that seem hard to justify without some prior understanding of what degrees of belief are.

In the pragmatic argument, the agent's credences in two propositions were said to be equal when the agent used them to govern acts that depend on those propositions in corresponding ways. Differences in utility were said to be equal when they correspond to the tradeoffs the agent was willing to make for the same degree of belief. This constructs a numerical scale of utility, which could then be used to construct a numerical scale for degree of belief, which was then shown to satisfy the probability axioms. All the numbers involved had a definite grounding in the behavior of the agent. But for the subject of alethic evaluation, it's less clear how to choose which numbers to use to represent degrees of belief. I will return to this question in Section 4.1.3.3, where I suggest that the best approach to this for the alethic argument is given by Lindley (1982) and Ajalloeian (2025).

3.3.6 Generalizations of the argument

There have been various generalizations of this argument beyond Joyce's results. Predd et al. (2009) show that if the total inaccuracy of a doxastic state is given by the sum of the inaccuracies of the degrees of belief in individual propositions, and if the inaccuracies of the degrees of belief in each proposition are each given by a proper scoring rule (though each proposition could use its own proper scoring rule), then again, any set of degrees of belief that violate the probability axioms are inadmissible. Pettigrew (2022a) shows that for this admissibility argument, it is not necessary for the inaccuracy of a doxastic state to be decomposable as the sum of inaccuracies of the degrees of belief in individual propositions—but a version of propriety is still needed.

All the results so far have focused on doxastic states over finite sets of propositions. Many of the measures of inaccuracy (particularly the ones that are sums over the inaccuracies of individual propositions) are infinite when there are infinitely many propositions. Thus, it is hard to make sense of what it means for one doxastic state to be alethically better than another according to such a measure. Schervish et al. (2009) suggest, among other things, that we consider the inaccuracy of *parts* of the doxastic state. Even if the doxastic state as a whole always has infinite inaccuracy, if there is a violation of the probability axioms that involves only finitely many propositions, then there is an alternate set of degrees of belief for those propositions that is guaranteed to be less inaccurate in every possibility. But Kelley and Neth (2023) point out that this only justifies the finite parts of probabilism, rather than the infinite axiom of countable additivity (which will be discussed more in Section 5.3). However, Nielsen (2023) and Kelley (2023) show how to generalize these results to at least some of the infinite parts of probabilism.

All of the arguments for probabilism work by dominance: we show that if a doxastic state violates probabilism, then there is a different doxastic state that is guaranteed to be less inaccurate in every possibility. It seems clear that this would mean that the doxastic state is irrational. In the picture that we are considering in this chapter, the role of the doxastic state is to be accurate. If one doxastic state is more inaccurate than another in *each* doxastic possibility for a subject, then the subject takes the former to be more inaccurate than the latter—they may not know which doxastic possibility is actual, but in all of them the comparison goes the same way. But Richard Pettigrew has, in a series of papers and books, shown that various additional principles connecting inaccuracy in different possibilities to rationality could lead to stronger requirements than just probabilism on a doxastic state.

For instance, Leitgeb and Pettigrew (2010b) show that if a subject's degrees of belief in different possibilities are not equal, then there is another doxastic state that *does* have equal degrees of belief in each possibility such that the *sum* of the inaccuracies in all possibilities is lower. Thus, if there's a reason why subjects whose doxastic state aims to play the alethic role should care about the *sum* of the inaccuracies, and not just the inaccuracy in each possibility, this would produce an extra requirement of rationality. Pettigrew (2016b) has connected a different rationality principle to this same requirement for degrees of belief. Instead of considering the *sum* of the inaccuracies in different doxastic possibilities, he considers the *worst* inaccuracy in any possibility and shows that the doxastic state on which each possibility has equal degree of belief minimizes this worst inaccuracy. Pettigrew (2012) considers various "objective chance" functions and shows that violations of a principle known as the "Principal Principle" (Lewis, 1980) would mean that a different doxastic state would have lower *expected* inaccuracy according to all possible chance functions. These arguments are summarized, with slight modifications, in Pettigrew (2016a). In his later book, Pettigrew (2022b), he generalizes the earlier unique concern for the worst inaccuracy or equal additive concern for the inaccuracy in each possibility, and instead allows that there might be different weights of concern for inaccuracy in each possibility, and derives different constraints on degrees of belief.

I think it is far from clear that a doxastic subject's degrees of belief has any rational requirement to do well according to any of these types of evaluation, so it's far from clear that any of these extensions of probabilism is rationally required. In Section 3.6.1, I will argue that a doxastic subject may have reason to care about expected inaccuracy calculated according to their own degrees of belief, but this doesn't justify any of the principles that ground any of Pettigrew's further arguments. In Section 8.1, I will in fact argue that *no* principles beyond probabilism *could* be rationally required.

3.4 CHALLENGES TO THE ACCURACY-BASED ARGUMENT FOR PROBABILISM

However, there have also been a series of challenges to this program in recent years.

3.4.1 Credence-state dependence

Although Hilary Greaves did some important early work for this accuracy program (which I will discuss more in Section 6.3.4), she has pointed out (2013) that there are parallels to some of the challenges of practical decision theory that arise in the epistemic case. She has pointed out that when one considers propositions whose truth value might be affected by one's attitude toward them, we get analogs of the Newcomb problem that was discussed in Section 2.5, and the more complicated challenges to the versions of decision theory that are aimed at solving it.

She considers a case where a mountain climber is about to attempt to leap across a chasm. Greaves suggests that there may be some cases in which the climber's actual success is more likely if she is highly confident that she will succeed. Consider the climber's credence in the proposition p: "I will successfully leap across this chasm". Alethic evaluation of p involves comparing the climber's credence in p to the truth value of p. But the truth value of p depends on her credence in p in a way that undercuts some of the applications of the standard alethic evaluation. Michael Caie (2013) and Jennifer Carr (2017) show that in some cases *violation* of the probability axioms actually ends up being evaluated best! For instance if p is a proposition that will be true if the agent's credence in it is less than .5, then it turns out that the way for the agent to minimize inaccuracy is to have credence just under .5 in p—but also to have credence 0 in $\bar{p}$, violating probabilism.

Some might argue that the dependence of a proposition's truth value on the subject's degree of belief in it is a pathological feature that can reasonably be ignored when considering alethic evaluation of the doxastic state. But the fact that belief in a proposition can make it more or less likely to be true is one of the very issues that motivated William James in proposing this sort of truth-based evaluation of doxastic states! How much this is a challenge to the grounding of Bayesianism in the alethic role for doxastic states may depend on how widespread these sorts of propositions are.

3.4.2 The separateness of propositions

Another problem Greaves raises (which is also raised by Selim Berker (2013) and Caie (2020)) is that the constraints we derive on degree of belief only arise from evaluating the *total* closeness of a subject's *overall* doxastic state to the truth. This means that even if the subject's credence in one proposition doesn't affect that very proposition's truth value, it could have impacts on the overall alethic assessment of the doxastic state in other unwanted ways. Several of these authors put this in terms of "epistemic bribes"—for instance, a religious organization might provide large grants for scientific research that would enable the answering of many questions, but only to researchers with the same religious beliefs as this organization. Berker notes that on an alethic evaluation of one's doxastic state, adopting these religious beliefs would seem to do well. But he argues that this would still be epistemically improper. He argues that this is like the tradeoffs endorsed by some utilitarians and criticized by some deontological ethicists, where one is intentionally sacrificed for the benefit of many.

However, some defenders of the alethic role for the doxastic state have defended it from this objection. Sharadin (2018) argues that the ethical analogy is imperfect. In particular, while *persons* have interests separate from each other's, that might be wronged by sacrificing one for the benefit of others, the application of this analogy to the doxastic case would require us to see each particular attitude in a person's doxastic state as somehow having its own interests that can't be traded off against that of others. If we are really talking about the doxastic state of one person (as opposed to a group, or epistemic community), this may not be reasonable. And Singer (2019) argues that in fact we do accept analogous tradeoffs in many cases, not only for ourselves but for others. Not only do scientists form beliefs in the absence of evidence that enable them to pursue the experiments that lead to evidence (as William James noted), but we teach students theories that we know are incorrect (like Newtonian physics) expecting that this will lead the students to more knowledge in the long run (including eventual revisions to this incorrect belief).

But Hieronymi (2006) raises a deeper version of the worry, while aiming to address a different issue. She accepts that belief has a constitutive orientation towards the truth, as part of a discussion of whether we can control our beliefs. (I will return to this discussion in more detail in Section 3.4.4.) To control a doxastic state, one would have to be able to subject it to some sort of goal other than its own truth. But this is incompatible with the constitutive orientation to the truth. While this might seem good for the alethic picture, it raises problems akin to those of Berker (2013). Noting that the doxastic state *as a whole* is farther from the truth than a particular probabilistic state only serves as a criticism of the doxastic state as a whole—it is not a problem for any *particular* degree of belief in it. If you look back at Figure 3.3, you can see that when the degrees of belief in p and in $\bar{p}$ add up to more than 1, the doxastic states that dominate it all differ in *both* degrees of belief. This turns out to be a general feature on all the proper scoring rules—fixing violations of additivity will generally involve changing the degrees of belief of all the propositions involved, and whatever the truth is, some of them will be moving closer to the truth while others will be moving farther away. On Hieronymi's picture, controlling one's degrees of belief in a way that ensured that they satisfy probabilism would require making the degrees of belief sensitive not just to their own accuracy, but to that of the others as well, and this would violate their constitutive aim.

Kolodny (2007) argues that coherence norms of any sort (he primarily discusses consistency for beliefs, but his argument applies just as well to probabilism for degrees of belief) are actually better explained by evidential norms. Although inconsistent beliefs can't both be true, the criticism of someone with inconsistent beliefs is not that their beliefs aren't all true (after all, one of their beliefs *is* in fact usually true!), but rather that we can tell that at least one of their beliefs is not properly supported by the evidence. (This requires further support for the claim that evidence can never support each of an inconsistent set of beliefs.) Similarly, one might say that although non-probabilistic credences are collectively farther from the truth than they have to be, one can factor the problem into a problem for individual degrees of belief by claiming that at least one of the individual degrees of belief is incorrectly responsive to the evidence. (Again, this requires further support for the claim that evidence can never support each of a non-probabilistic set of degrees of belief.) I suspect something like this idea is part of Berker's deeper motivation and perhaps that of some of the others. I will discuss the prospects for this sort of argument further in Section 3.5.

3.4.3 Propriety

Another recurring challenge to the accuracy program has concerned the role of propriety as a feature of the measure of inaccuracy. Both Blackwell and Drucker (2019) and Daoust and Côté-Bouchard (2023) raise some of these worries. Although there are a variety of arguments for probabilism, based on a variety of different constraints on the scoring rule, all of them require propriety, either explicitly or implicitly. As a reminder, in the situation where the overall measure of inaccuracy factors into the functions $I(x,0)$ and $I(x,1)$ measuring the inaccuracy of a given credence in each proposition when it is true and when it is false, it takes the form that

$$xI(x,1) + (1-x)I(x,0) < xI(y,1) + (1-x)I(y,0),$$

when $x \neq y$. When the alethic value is thought of as a feature of the doxastic state as a whole, rather than of the individual degrees of belief in individual propositions, the requirement is that for any probability function, the expected alethic value of any doxastic state other than the one described by that probability function is worse than the expected alethic value that probability function calculates for itself.

Historically, propriety originated as a condition when scoring rules were used to set pragmatic payouts for weather forecasters (Brier, 1950) or in other contexts where statisticians wanted to accurately elicit the degrees of belief of someone motivated pragmatically (de Finetti, 1974). Since the person could be assumed to already have a set of degrees of belief that play the pragmatic role, the person could be assumed to make decisions in a way that maximizes their own expected pragmatic value. By setting up a payment schedule for the person's reports that satisfies propriety, one could ensure that the person would be incentivized to report their actual degrees of belief, rather than some other function of them.

This motivation doesn't obviously work for a pure alethic evaluation of a doxastic state, rather than an imposed pragmatic evaluation of it. But there may be an argument of a similar sort nevertheless. In Section 3.6.1, I will argue that there may be a way in which a subject who has a doxastic state that plays the alethic role is motivated to govern their actions in line with expected value. If this is right, then for a subject whose alethic value is governed by an improper scoring rule, certain probability functions would be unstable. Someone whose doxastic state gave higher expected value to a different doxastic state than their own would be motivated to change their doxastic state. If we think that no probability function should be unstable in this way, then that would rule out improper scoring rules as appropriate ways to evaluate alethic value. But why should no probability function be unstable in this way? In the context of an argument for probabilism, this assumption seems unacceptably question-begging.

Joyce (1998, 2009) gives the argument that every probability function could give the objective chances of some part of the world, and that the alethic evaluation of a doxastic state shouldn't prevent the doxastic state from stably matching the objective chances. But Blackwell and Drucker (2019) argue that in fact *not* every probability function could give the objective chances. And in any case, the argument depends on the success of something like Pettigrew (2012)'s extension of the accuracy argument to support not just probabilism, but matching the objective chances. (And it also depends on the *existence* of objective chances, which some Bayesians have found questionable, at least since de Finetti (1974).)

In the case where the alethic evaluation of a doxastic state can be factored into the alethic evaluation of the degree of belief in each individual proposition though, I think the argument can be strengthened. If the functions $I(x, 0)$ and $I(x, 1)$ didn't satisfy propriety, then not only would there be certain *probability functions* that were unstable—there would in fact be specific *degrees of belief* that were unstable. If

$$(1/3)I(1/3, 1) + (2/3)I(1/3, 0) > (1/3)I(1/4, 1) + (2/3)I(1/4, 0),$$

then anyone who had degree of belief 1/3 in some proposition would expect degree of belief 1/4 in that proposition to be less inaccurate. We established a one-dimensional numerical representation of degrees of belief evaluated alethically by eliminating all hypothetical degrees of belief that are worse than one of these numerical ones both when the proposition is true and when it is false. This numerical representation then ensured that $I(x, 0)$ was strictly increasing and $I(x, 1)$ was strictly decreasing, when $0 < x < 1$. Violations of propriety would rule out particular numerical degrees of belief within this scale as well. It doesn't require a commitment to probability functions to think that this is problematic.

Still, I will suggest in Section 4.1.3.3 that the best response should not be committed to a specific numerical representation of degrees of belief, the way probabilism is. Instead, just like in the pragmatic case, we should see the specific choice of numbers for the arbitrary commitment that they are. The inaccuracy measure can be given by any continuous pair of an increasing and decreasing function. This will give rise to some set of numerically-represented doxastic states that are admissible. And as Lindley (1982) shows (and Ajalloeian (2025) puts into a proper normative context), there will always be some way to transform the numerical scale in such a way that the admissible doxastic states turn out to be precisely the ones that satisfy the probability axioms. Once this transformation has been carried out, the original inaccuracy measure will have been transformed into a proper scoring rule.

3.4.4 Voluntarism

Another question that sometimes arises in discussion of the alethic program is that of whether the doxastic state is under the subject's control. The motivation for propriety, namely that impropriety makes certain doxastic states unstable, seems to depend on the idea that the subject would *change* their degrees of belief if they expected some other degrees of belief to be less inaccurate. Even the force of inadmissibility, thought of as a *reason* not to have a non-probabilistic doxastic state, seems to suggest that the doxastic state is under the subject's control in some way.

However, there has been a long debate in the history of philosophy over the extent to which belief is or isn't under a person's control. I won't rehearse the full debate, but details are available from Boespflug (2024). It seems obvious to most of us that we can't just choose to believe that there are currently an odd number of people in the Empire State Building, and that we can't just choose not to believe that the sky is blue. But others have claimed that there are propositions that we can choose to believe or not, whether it's Descartes thinking that *all* our errors are the result of our own faulty choices of what to believe, or someone who chooses to believe that they closed the garage before leaving for vacation, as a way to settle their worries about whether they might have left it open. Some of these thoughts are about whether, as a contingent matter of fact, humans can't actively control our doxastic states, while others argue that this is conceptually central to the concept of something as a doxastic state.

To some extent, the question of voluntarism isn't actually essential to Bayesianism. Both pragmatic and alethic arguments for probabilism (and other aspects of Bayesianism to be discussed in Chapter 6 and later) work without it. The pragmatic argument for probabilism shows that, if the doxastic state is a feature of an agent that they use to guide their actions in particular ways, then violations of probabilism mean that there is a particular sort of way that their doxastic state will guide them badly. The alethic argument for probabilism shows that, if the doxastic state is a feature of a subject that has a role of closeness to the truth, then violations of probabilism mean that there are particular alternative doxastic states that would do the job better—their doxastic state is thus doing a bad job. If the doxastic state is voluntary, then perhaps both of these arguments have deontic force—the subject should do better. But even if the doxastic state is just a feature that the subject themself does not control (as it clearly is for the neural nets considered above, whether or not it is for us), there is still a sense in which it is unfortunate that it does its job badly. There may be a creator or designer to blame or we may just say that this is one of the tragedies of existence, that happens to afflict this subject. But the subjects that do their job well will be (roughly) Bayesian. Rinard (2017) argues that this is sufficient to say that the requirement of Bayesianism is a kind of rational requirement, but I think it doesn't matter too much whether we use that word in that way or not.

However, some conceptual objections to voluntarism may go deeper as criticisms of the alethic role for the doxastic state. The alethic role for the doxastic state is often put in the slogan that "belief aims at the truth". But Vermaire (2025) argues that belief, and in fact any doxastic state, can't *aim* at all, and since anything that is controlled is controlled with an aim, doxastic states can't be controlled. For our purposes, the argument that doxastic states can't aim is of greatest relevance. He argues that aiming at a goal requires being guided by a doxastic state that represents one's means as a means to that goal. One doesn't need to *believe* that the means will be effective, but *some* doxastic state is necessary—one can't aim at winning the lottery by buying a ticket without some degree of belief that buying the ticket might lead to winning. This calls into question the very idea of a creature whose doxastic states are constitutively *aimed* at the truth—the neural nets described above might not count as doxastic subjects for him, but just as tools of the researcher that creates them. But this also rejects the central idea of James (1896).

3.5 TRUTH VS. EVIDENCE

One natural response to these issues is to follow Clifford (1877) and say that beliefs rationally ought to be controlled by the evidence, rather than aiming at accuracy. This idea has been traditional in much epistemology since Plato's claim that knowledge is better than mere true opinion because it is in some sense "tied" to the truth. By the 20th century, this came to be described in terms of beliefs being "justified". Since Gettier (1963) (who describes cases in which true beliefs are justified by evidence that turns out to be misleading and unconnected to the actual reason the propositions are true), many epistemologists have sought to characterize what connection between the truth of a proposition, one's justification for believing it, and one's belief must hold in order for the belief to qualify as knowledge.

In my (2017) I argued that we might think about truth, evidence, and action as three separate handles we can get on doxastic states. Action lets us grab on because it is what belief causes. Truth lets us grab on because it is what belief is supposed to be. And evidence lets us grab on because it is what causes belief. However, there has been relatively little work that uses the evidential role directly to give an understanding of how degree of belief should behave. To some extent, this is because there has been so much success at using the pragmatic and alethic roles for a doxastic state to justify what appear to be evidential norms, in terms of conditionalization on the evidence. I will discuss these arguments in Chapter 6. But if the appeal to the evidential role is meant to avoid various worries for the pragmatic or alethic role, then an independent characterization is needed. Unfortunately, no independent characterization of the evidential role has been worked out in anything like the detail that the pragmatic and alethic roles have been. (I give a brief gesture toward some such independent characterization in Section 4.1.3.4, but it is definitely not worked out.)

However, there have been a few hints. Recall that Skyrms (1984) and Christensen (1996) somewhat anticipated Joyce's alethic arguments for probabilism. They were somewhat unsatisfied with the pragmatic argument, and suggested that the pragmatic arguments might merely be a tool for diagnosing some more truly epistemic failing in non-probabilistic degrees of belief. In fact, Joyce himself has sometimes suggested a parallel thing about his accuracy arguments, in his (2005) and in various unpublished remarks. Skyrms and Christensen can be understood as making something like the following argument:

- Good degrees of belief would never govern actions inconsistently.
- Violations of the pragmatic axioms result in inconsistent actions.
- Therefore, degrees of belief that violate probabilism can never be good.

This doesn't yet turn the pragmatic argument for probabilism into an alethic or an evidential one. In particular, it requires characterizing "good degrees of belief" in this independent way, and then justifying the claim that they would never govern actions inconsistently. (I will consider this in Section 3.6.1.) Similarly, Joyce can be read as making something like this argument:

- Evidence would never justify degrees of belief that are alethically dominated.
- Non-probabilistic degrees of belief are alethically dominated.
- Therefore, non-probabilistic degrees of belief would never be justified by the evidence.

Filling this argument out requires giving a more detailed characterization of how evidence justifies degrees of belief, and why it wouldn't justify alethically dominated degrees of belief (or, indeed, ones that govern actions inconsistently).

Pettigrew (2014) considers some other possibilities for how alethic and evidential value might interact. If evidential and alethic value are both fundamental aspects of the evaluation of a doxastic state, there are various problems. As observed by Easwaran and Fitelson (2012), if evidence functions by putting certain *constraints* on a doxastic state, but accuracy is then used to evaluate the remaining doxastic states, then Joyce's argument for probabilism fails. A doxastic state eliminated as "inadmissible" because some other possible doxastic state was alethically better than it in all possibilities may not be eliminated if all the dominating states were ruled out by the evidential constraint. For the argument to work, either accuracy considerations have to underlie the evidential constraints (as Pettigrew eventually argues, pointing to the work I discussed in Section 3.3.6) or accuracy considerations have to be merely diagnostic of some underlying evidential issue (as Joyce suggests).

3.6 USING THE ALETHIC OR PRAGMATIC ROLE TO GROUND THE OTHER

Just as we might consider how alethic considerations could ground the use of evidence, or evidential considerations could ground the alethic arguments, it is worth going back to consider how either the alethic or pragmatic roles for a doxastic state could ground the other. That is, if a creature has a kind of state that aims to represent the world with the function of being close to the truth, we would like to show that such a creature would also have reason to use this state to guide its actions in ways that line up with those of creatures whose doxastic state has the primary function of guiding actions. Similarly, if a creature has a kind of state that aims to represent the world with the function of guiding its actions effectively, we would like to show that such a creature would also have reason to evaluate the accuracy of this state in terms of closeness to the truth. The fact that both of these projects can be done, as I will soon show, goes a long way toward establishing that both of these are ways for a creature to count as having a doxastic state.

3.6.1 From alethic to pragmatic

My (2014a) shows how a doxastic state that is constitutively aimed at the truth might nevertheless be used to rationalize action, for a creature that also has an independently existing attitude of utility, or something similar. Without some sort of bouletic state like this, the creature would have no interest in guiding their actions. Just like Savage, I consider the creature's preferences over all hypothetical actions, considered as functions from states to consequences. I don't know if something like my procedure could be used to ground something like evidential, causal, or functional decision theory.

The basic idea is that I argue that preference should at least be a partial ordering of acts, which we can represent by $\succ$, with the condition that if $A \succ B$ and $B \succ C$, then $A \succ C$, and that $A \not\succ A$. I start with a very minimal ordering, and then show how the doxastic state of the subject could rationalize certain extensions of the ordering.

I begin with the idea of dominance. If on every state, act A yields a consequence that is strictly better than the one yielded by act B, then $A \succ B$, and if it is better on some states and equal on others, then at least we can say that $A \succeq B$. (I allow that it might also be the case that $A \succ B$ in such a case, but I claim that it might follow from some further principle rather than just from dominance.) So far, the subject's doxastic state does not play a role.

However, the doxastic state starts to become relevant when considering acts that don't directly bear a dominance relation. If the subject's doxastic state reflects their best guess about the truth, I claim that they can use it to guide their actions. In particular, if A and B are two acts, and if for all consequences x, the subject has the same credence in proposition that A yields an consequence at least as good as x as they do that B does, then they think of the value of the consequences of these acts the same way. In this case, I claim that the subject should be indifferent between the two acts, $A \approx B$. Putting this together with dominance, we see that the subject's preferences among acts should at least satisfy the principle of "stochastic dominance" endorsed by Tarsney (2020).

If consequences further have a scale on which we can determine equal differences, then we can get another rule for indifference that we might call "tradeoff equivalence". We should be indifferent between A and B if we have the same credence that A will be better than B as that B will be better than A, and if we're certain that the amount A might be better than B is exactly equal to the amount that B might be better than A. Ramsey (1926) used a pre-existing concept of when an agent was indifferent between two acts to determine a scale on utility; here I use a pre-existing scale of utility to determine when a subject should be indifferent between two acts.

I then show that, if degree of belief obeys the rules of probability (perhaps justified by alethic considerations), and if any set of states can always be further subdivided into a finer partition, then whenever the expected utility of A is higher than the expected utility of B, these three principles entail that $A \succ B$. Instead of deriving expected utility as a convention of representation of a preference ordering that is governed by its own rules, we derive it from assumptions about degree of belief and utility and the way that they ought to interact. Furthermore, we can still get some constraints on preference even if not all of the scale assumptions on utility can be justified. Furthermore, we can justify principles governing preference for various types of action where expected utilities are infinite or undefined.

A doxastic state whose primary function is to be accurate, may nevertheless be used to rationalize certain preferences over acts.

3.6.2 From pragmatic to alethic

Conversely, Levinstein (2017) shows how a doxastic state that is constitutively aimed at governing action might nevertheless be used to rationalize a particular measure of the inaccuracy of doxastic states. He notes that a doxastic agent has preferences over acts, and also has preferences over outcomes. They don't directly care about what degrees of belief they have or how close they are to the truth.

However, if their degrees of belief include degrees of belief about what decision problems they might face, then something like a strictly proper scoring rule can be derived. Levinstein notes that an agent that has a sense of what decisions they will face, and what the consequences of those decisions would be in a world, can use this to form an estimate of the goodness or badness having a given doxastic state in that world.

First, he considers what decision problems they might face. He starts by considering the possibility that they will face a choice about whether or not to buy a bet on a given proposition at price q. In a world where the proposition is false, they lose q if they buy this bet, compared to the alternative of not buying it. In a world where the proposition is false, they lose $1 - q$ if they fail to buy this bet, compared to the alternative of buying it. He represents the consequences of these decision problems entirely in terms of the hypothetical loss the agent would face compared to the alternative, rather than in terms of gains.

If the agent's degree of belief in p is x, they will buy the bet if $q < x$ and they will not buy it if $q > x$. Thus, to find their expected loss from facing such a bet, we should consider their degree of belief that they will face the choice of buying a bet at price q on such a proposition, and integrate q times this degree of belief for $q < x$ if the proposition is false, and integrate $(1 - q)$ times this degree of belief for $q > x$ if the proposition is true.

He then cites a theorem of Schervish (1989) to show that the result of this integration will always be a proper scoring rule—and in fact, every proper scoring rule arises this way for some distribution of degrees of belief over the q that they will face. He thus shows that a pragmatic agent is able to make use of the results justified by alethic evaluation of doxastic states—and can answer the central question of why to use a proper scoring rule, and also which one.

A doxastic state whose primary function is to govern preferences over acts, may nevertheless be used to rationalize a particular degree of concern for accuracy.

3.7 CONCLUSION

The distinctive feature of an alethic subject is that they have a doxastic state that aims to be close to the truth. Some epistemologists have argued that this sort of idea must be closer to the core of epistemology than doxastic states with a purely practical role, but it turns out that many of the features that the two roles can justify end up being basically the same. But it is worth distinguishing the alethic role from an evidential role, where the connections have not yet been worked out in as great detail.

If belief plays the alethic role, while degree of belief plays the pragmatic role, we can justify the Lockean thesis. If belief plays the alethic role, and there is no independent concept of degree of belief, then we can still use belief to ground a concept of degree of belief. But much of the most interesting work on the alethic role concerns subjects that just have degrees of belief.

There are some questions remaining about how to get started in grounding degree of belief in the alethic role. There must be some justification for the use of a numerical scale for accuracy. Once this exists, we can argue that degrees of belief must be representable by real numbers on the interval from 0 to 1, but there are still questions about why to choose one particular numerical representation over another. For various reasons, authors in this literature have assumed that the measure of accuracy must be a proper scoring rule, but several of these reasons are better motivated on a pragmatic account than a purely alethic one. Despite these questions about the foundations, there are some good examples of subjects that are best understood in this way, namely certain kinds of simple neural nets.

If accuracy is measured by the Brier score, then a simple geometric argument shows that degrees of belief that violate probabilism do a bad job of being accurate. Various authors have extended this argument to various classes of proper scoring rules. There are some difficulties extending this argument to the full consequences of probabilism when there are infinitely many propositions. But if there are motivations for decision rules beyond dominance, then we might be able to use accuracy to justify more than just probabilism.

There are some important worries for the foundations of the alethic arguments. The problem of act-state dependence is potentially more significant for the alethic justification than for the pragmatic justification, though it is not fully solved for either. There are also conceptual worries about evaluating individual degrees of belief in terms of their contribution to an alethic evaluation of an overall doxastic state, rather than separately. There are worries about propriety. And there are worries that the entire framework may depend on a conception of degree of belief as something that is under the subject's control in a problematic way.

Some epistemologists would rather ground the doxastic state in an evidential role, but the details of this have not been worked out yet. However, the alethic and pragmatic roles can each be argued to justify the core commitments of the other. Thus, there is a good case to be made that these are both ways that a creature could count as doxastic, and I claim, could ground what matters in epistemology.

RESEARCH QUESTIONS

1. Can there be purely alethic value independent of any practical value?
2. Relatedly, can there be a subject that counts as doxastic without having any practical desires?
3. Can the numerical measurement of alethic value be justified?
4. Can there be a purely alethic grounding of the relevant numerical features of degree of belief or alethic value?
5. Just as an agent with solely practical concerns can be led to instrumentally care about alethic value, can a subject with solely alethic concerns be led to instrumentally care about certain practical values?
6. Can the role of evidence be derived purely from considerations of accuracy or pragmatics?
7. Is there a formal theory of evidential value that isn't derived from accuracy or pragmatic value?

CHAPTER 4

The measurement of belief

CHAPTER 1 OF this book made the case that there is a useful role in epistemology for an attitude of degree of belief that can be higher or lower. Chapters 2 and 3 considered how we might understand this attitude's role either in governing action or in representing the world truthfully. Both of these roles seem to lead to the idea that degrees of belief come on a numerical scale satisfying the probability axioms. That is, for any propositions p and q:

- $P(p) \geq 0$.
- If the subject is certain of p, then $P(p) = 1$.
- If the subject is certain that p and q aren't both true (i.e., the propositions are disjoint sets of possibilities), then $P(p \cup q) = P(p) + P(q)$.

However, many philosophers have claimed that this sort of probabilism requires unrealistic precision and numbers in the head, and therefore it can't be correct either as a theory of actual people or even of the sorts of models we might aspire to emulate. In this chapter I will consider several of these objections in order to defend the idea of degrees of belief as precise, numerical probabilities. Although I don't have positive evidence from empirical psychology to suggest that actual humans really do behave this way, I will argue that the sorts of features philosophers usually point to in arguing against precise numerical probabilism are actually quite compatible with it. Whether or not we in fact have

DOI: 10.4324/9781315626918-4

this sort of degree of belief, creatures that do have it wouldn't have to be notably different from us.

This discussion will also be useful in clearing up some misconceptions about what it could mean for degrees of belief to be precise numerical probabilities. I will use many tools from the theory of measurement in social science and the philosophy of science to compare numerical degrees of belief to many other features of the world that we feel happy representing numerically. Probabilism itself will be seen to be something like the Fahrenheit scale for temperature—it can accurately represent the underlying facts in a precise and numerical way, even though many aspects of it are pure conventions that could have been represented otherwise.

There are three main families of objections to precise numerical probabilism that I will address in this chapter. The first set of objections focuses on the structure of the arguments from the previous chapters in favor of probabilism, claiming that these arguments couldn't tell us about degree of belief itself, but only about how we choose to represent it. The second set of objections claims that degree of belief (or any related notion of "partial belief") could at most explain a small amount of the psychological behaviors that are claimed for it. The third set of objections claim that in any case, any numerical measurement of degrees of belief could at best be extremely imprecise, so that using precise real numbers is drastically misleading.

4.1 REPRESENTATIONALISM

The first family of objections is exemplified by Zynda (2000) and Meacham and Weisberg (2011). They focus on the representation theorem arguments (discussed in Chapter 2), and claim that these can't establish the literal truth of the claim that degrees of belief either do or should satisfy the axioms of probabilism. At most, they claim, these arguments can serve a sort of heuristic purpose. I will respond by clarifying the role of numbers in the probabilistic representation of degrees of belief. I will argue that the challenges raised by these authors is no more significant than analogous challenges we can raise to the role of numbers in our thinking about temperature, or distance, or many other physical quantities. While their challenges don't obviously apply to the accuracy arguments discussed in Chapter 3, I think that considering these arguments can help clarify our thinking about the role of numbers there as well, and what we must assume about mental representation in order for that argument to justify probabilism.

Zynda phrases the representation theorem argument for probabilism as using the following three premises:

> **Representability.** If a person's preferences obey the axioms of expected utility theory, then he or she can be represented as having degrees of belief that obey the laws of the probability calculus. ...
>
> **The Rationality Condition.** The axioms of expected utility theory are the axioms of rational preference.
>
> **The Reality Condition.** If a person can be represented as having degrees of belief that obey the probability calculus, then the person really has degrees of belief ... that obey the laws of the probability calculus. (p. 51)

Together, these premises lead to the conclusion that a rational person really has degrees of belief that obey the laws of the probability calculus. As we saw in Chapter 2, there are significant questions about which particular axioms of expected utility theory are requirements of rational preference. (At least, Ramsey, Savage, Buchak, Jeffrey, and Joyce all give different such axioms.) One might worry that the correct axioms of rational preference only suffice to prove a weakened form of Representability. However, Zynda grants that we might be able to find an appropriate combination of axioms and theorems, and instead focuses on the Reality Condition. Even if we can ensure that a rational person can be *represented* with a set of degrees of belief that obey the probability calculus, he thinks there may be no sense in which they really *have* such degrees of belief.

4.1.1 Four views

He considers four different views that one might have about degrees of belief, borrowing terminology from philosophy of science and philosophy of mind. He calls these views "eliminativism", "antirealism", "weak realism", and "strong realism". These four views are a useful framework for understanding what we think of any putative category in the world. The focus of his argument is that a theorist who relies on the representation theorem argument for probabilism must reject strong realism. Meacham and Weisberg go further and say that such a theorist must reject even weak realism. My aim is to defend the idea that strong realism is compatible with all the challenges raised by Zynda, and Meacham and Weisberg.

4.1.1.1 Eliminativism

Eliminativism about a concept is the view that this concept doesn't correspond to anything in the world. This is the view that contemporary scientists take to historical concepts like "phlogiston". Early modern alchemists had developed the historical elemental theory of earth, air, fire, and water, and had come to the conclusion that there were multiple separate elemental earths, and thought of fire as the emission of a substance called "phlogiston" that is part of the composition of various fuels. Through their understanding of the smelting process, they had come to the conclusion that the different metals (copper, iron, silver, gold) were each compounds of a particular earth with phlogiston. They also recognized phlogiston as a substance given off in the respiration of animals. The English scientist Joseph Priestley was able to extract a form of air he thought of as "dephlogisticated", because it promoted combustion and helped the respiration of animals. His work was translated into French by Marie-Anne Paulze Lavoisier, who assisted her husband Antoine-Laurent Lavoisier in various experiments proving that metals actually *gain* mass when converted to the corresponding earth. They argued that Priestley's "dephlogisticated air" was actually a new element, which they called "oxygen", and that combustion and respiration should be understood in terms of absorption of oxygen, rather than emission of phlogiston. In contemporary theory, the different roles played by phlogiston in alchemical theory are played in various places by spare electrons, absence of oxygen, reducing potential, and various other concepts, but phlogiston itself has been rejected. There are similar histories with the "caloric" theory of heat as a fluid (rejected in favor of a statistical mechanical explanation in terms of the energy states of the microscopic constituents of a substance), and the "luminiferous ether" that supposedly conducted light (rejected by Einstein's theory of relativity, in which there is no fixed substance that electromagnetic waves excite).

Eliminativism has been an important view in the philosophy of mind, strongly associated with the work of Paul and Patricia Churchland (Churchland, 1981, 1986), but with important precedents in the work of Wilfrid Sellars (1956) and others. The idea is that just as the concept of phlogiston turned out to be a mistake (albeit a useful one) on the way toward understanding how substances really transform, our ordinary psychological concepts like belief, desire, and perception are also mistakes on the way to a better understanding of the mind.

Zynda thinks that the eliminativist about mental content is unreasonable, because he or she denies the existence of "those features that contribute essentially toward making him or her a person and agent". However, I think thoroughgoing eliminativism about mental concepts generally is a more serious challenge—perhaps personhood and agency can be reinterpreted in the new terms (as combustion was reinterpreted in the oxygen theory) or perhaps we should reject those concepts too once we better understand our nature as physical animals. I think that properly responding to the eliminativist requires a deeper empirical study of the human mind and behavior than I am in a position to carry out. For further discussion of eliminativism, see Ramsey (2013).

In any case, the main thrust of the eliminativist challenge is about actual humans, and as I argued in Chapter 1, I think the project of epistemology is meant to apply not just to actual humans but to other potential cognitive architectures that could count as doxastic as well. In the rest of this chapter I aim to respond to the sorts of challenges raised to the numerical measurement of degrees of belief by philosophers who don't aim to fully eliminate the concepts of epistemology from the human mind. I claim that the features of human cognition focused on by these philosophers are not unlike what one might expect from Bayesian subjects. Thus, unless one wants to reject the application of all ordinary concepts of epistemology to humans, I don't think there is clear reason to reject the application of numerical degrees of belief satisfying a version of probabilism.

4.1.1.2 Antirealism

The second view of a concept that Zynda considers is antirealism. Antirealism is sometimes taken as a global view about nearly *all* of the concepts of science. This view is perhaps most fully defended by van Fraassen (1980). On van Fraassen's view, the most our evidence tells us about the world is that our observations proceed *as if* there were subatomic particles and distant galaxies, and *as if* DNA were a molecule obeying particular chemical laws that guided the evolution of life over millions of generations. Since we can't directly observe the microscopic or the extremely distant or the far reaches of biological history, we can't be sure that there is anything real in the world corresponding to these theoretical concepts. At most, we can say that these theories correctly predict the outcomes of experiments.

This sort of global antirealism about theoretical concepts would be no special challenge to probabilism about degrees of belief. But some scientific theories have involved a more local antirealism. An example is the theory of "quasiparticles" in solid state physics. A substance like copper that is a good electrical conductor can be understood as one in which electrons have a very large number of allowable states at many energy levels, so that they can move very easily between them under the influence of electrical forces. A substance like rubber that is a good electrical insulator can be understood as one in which electrons have very few allowable states, so that they basically can't move at all. A semiconductor like silicon is a substance in which there is a set of low-energy states that electrons can easily move between, and also a set of high-energy states that electrons can easily move between, but it is difficult to get an electron to move between the low-energy and high-energy states. Ordinarily, all the low-energy states are filled, and all the high-energy states are empty, and the negative charges of these low energy electrons are balanced with the positive charges of the atoms. But if you introduce an extra electron into one of the high-energy states, then it can flow easily through the silicon as a negatively charged particle, without interference from the sea of electrons in the low-energy states.[1] Interestingly, if you instead *remove* an electron from one of the low-energy states, this allows nearby electrons to move easily to fill the hole. Instead of thinking about the slight movement of lots of electrons, it becomes easier to think of the *hole* as a "quasiparticle" that moves through the silicon as if it were a positively charged particle. Although only the electrons are *really* there, it's sometimes easier to talk about what's going on by talking about the holes.

Interestingly, this theory of holes initially arose not in the discussion of semiconductors but as the English physicist Paul Dirac's explanation of antimatter. He noted that if we pretend that the vacuum of space has this similar separation between high-energy and low-energy states for electrons, and pretend that there are electrons filling all the low-energy states, then we could observe holes moving as positive charges just as electrons move as negative charges. If an electron encounters

[1]More accurately, there *is* interference from the sea of electrons in low-energy states, but the net effect of this interference is that the high-energy electrons moves *as if* it were an electron of higher mass in empty space. This is another antirealist aspect of the treatment of semiconductors.

a hole, it can "fall down" to the lower energy state, releasing energy, and destroying both the free electron and the hole. This behavior of holes is exactly the observed behavior of positrons—the anti-particle corresponding to electrons. In semiconductor physics it is common to think of the hole as a fictional entity that simplifies our discussion of the behavior of the real low-energy electrons in different types of conduction. In vacuum physics it is common to think of the sea of electrons and the excess low-energy states they fill as a fiction that can help our thinking of the real antimatter. Van Fraassen would insist that all ways of thinking about the situation are equally fictional, while a local antirealist would suggest that one picture is the real picture, and the other is a useful fiction.

The difference between antirealism and eliminativism is that the eliminativist says that the concept is confused and leads to errors, while the antirealist allows that although the concept corresponds to nothing real in the world, it allows one to make the same predictions and explanations as the correct theory.[2] Antirealism in this sense is not so common in epistemology and the philosophy of mind, though the term "antirealism" is sometimes used to mean what Zynda and I call eliminativism. However, as I discussed in Section 1.5.2.3, in Easwaran (2015), I showed how a particular realist view of the concept of full belief could underpin an antirealist story of this sort about Bayesian degrees of belief satisfying the probability axioms. I didn't give any reason to think this sort of view would be true, but aimed to use it to show that a realist about full belief should not be a total eliminativist about the Bayesian view.

4.1.1.3 Weak and strong realism

Zynda's main focus however is the distinction between what he calls "strong realism" and "weak realism". While these views agree that the concept involved corresponds to something real, they disagree about how fundamental the concept is. A biological population is a real thing, but

[2] Depending on how detailed the predictions and explanations required are, these two views may shade into one another. The treatment of fuels as highly phlogisticated substances (rather than their modern treatment as reducing agents that combine with oxidizers) may be sufficient for some purposes that aren't too sophisticated. Meanwhile, the treatment of a hole as a quasiparticle behaving like a positively charged electron in a semiconductor does run into some difficulties near the edges of the solid, where it might have to interact with electrons from a substance with different energy levels.

it is composed of many individuals that compete and reproduce with each other. The individuals are more fundamental than the biological population, but there are useful laws that can be expressed at each level. The planets of the solar system are real objects, but they are composed of chunks of solid and gas (and in a few cases liquid) that move around in ways that are more fundamental. As Zynda points out, for some purposes we can treat the center of mass of the Jupiter system as the object that orbits the sun, but while this center of mass is real, its behavior is in some fundamental sense to be explained by the behavior of all the gases and metals and ices of the planet and its moons.

The distinction between strong and weak realism is comparative here. Neither Zynda nor anyone else (except perhaps some sort of Cartesian dualist about the mind) thinks that degrees of belief are fundamental irreducible entities in the world. Everyone agrees that degrees of belief, if they really exist, are somehow constituted (at least in humans) by the behavior of bodies and neurons, and their interaction with the world. Like anything else psychological or social, the extent of the physical world that constitutes these states is unclear. There is such a thing as the exchange rate between the US dollar and the Japanese yen, but is it just constituted by the attitudes of central bankers, or does it depend on the behavior of all international commerce between the United States and Japan, or does it even depend in minute ways on the degree to which I am prepared to act on an idle fantasy of purchasing a ticket to hike up Mt. Fuji? Similarly Putnam (1975) argues that the contents of a person's mental states depends not just on the states of her own body but also on her history of interaction with the objects her thoughts are about, and her tendencies to defer to various social 'experts' on the meanings of her concepts. Furthermore, Clark and Chalmers (1998) argue that one's mental state may partly be constituted by the various physical and social structures that shape one's mental life, from notepads to smartphones to communities.

But the question for Zynda is whether the preference that the axioms of a representation theory apply to is more or less fundamental than the degrees of belief that form part of the representation that is proved to exist. The strong realist about degrees of belief will say that, however degrees of belief are constituted in the world, they are part of what gives rise to the preferences of an agent. The weak realist, on the other hand, will say that preferences are the more fundamental mental state, and that degrees of belief, and perhaps even belief and desire themselves, are just useful summaries of a more fundamental notion grounded in something like our choice behavior.

I take it that something like the strong realist view is the one most people land on, once they accept the notion of degree of belief as something real. It certainly seems natural to say that one prefers one act to another *because* one is more confident that it will lead to a good outcome. But Zynda, like the philosophers and economists who work with representation theorems discussed in Section 2.6, wants to argue that this is an illusion. The only way that they say we can make sense of degrees of belief is as something implicitly defined by our preferences, rather than as something more fundamental that explains our preferences.

4.1.2 Zynda's challenge

Zynda argues that the strong realist can't support the Reality Condition of the representation theorem argument. To do this, he describes two agents, called Leonard and Maurice. Leonard and Maurice have all the same preferences over gambles—they both prefer to receive a prize if a fair coin lands heads rather than if a fair die rolls a 3, and both prefer to receive a prize if a fair die comes up less than 5 than if a fair coin lands heads.

However, they report their degrees of belief quite differently. Leonard reports his on a scale from 0 to 1, and says his credence that a fair coin lands heads is .5, while his credence that a fair die rolls a 3 is 1/6. His degrees of belief satisfy the probability axioms—they are non-negative, he has credence 1 in things he is certain of, and his credence in a disjunction is the sum of his credences in the disjuncts, if he thinks they are incompatible. Maurice, however, reports his degrees of belief on a scale from 1 to 10, and says his credence that a fair coin lands heads is 5.5, while his credence that a fair die rolls a 3 is 2.5. His degrees of belief satisfy a different set of axioms—they are at least 1, he has credence 10 in things he is certain of, and his credence in a disjunction is 1 less than the sum of his credences in the disjuncts, if he thinks they are incompatible.

Because these agents have the same preferences, they both satisfy the Rationality Condition of the representation theorem argument. Thus, by the Representation Theorem, both can be represented as having degrees of belief satisfying the probability axioms. In order to save the Reality Condition, therefore, we must deny the significance of Maurice's self-reports of violating the probability axioms. A weak realist might say that since degrees of belief are dependent on preferences, the self-reports don't matter, and the degrees of belief of both agents really are those reported by Leonard. But a strong realist must say that the degrees of belief are in some sense prior to the preferences. In order to save the Reality Condition, a strong realist must therefore say that whatever it means to "obey the probability calculus" is something that is shared by Leonard and Maurice. In particular, none of the numerical features of probability are literally true.

Meacham and Weisberg (2011) argue that this sort of view would deny too much of probabilism to be worth saving as a theory of real degrees of belief. They note that while Leonard and Maurice's self-reports differ just by a simple transformation of adding 1 and multiplying by 9, one could proliferate examples with other mathematical transformations. If someone else had degrees of belief that were the squares of those of Leonard, then this person would have credence 1/36 that a fair die rolls a 3 (where Leonard has 1/6), and 1/4 that a fair coin comes up heads (where Leonard has 1/2), and 25/36 that a fair die rolls less than a 6 (where Leonard has 5/6). Both people agree that rolling a 3 is least likely, getting heads is in the middle, and rolling less than a 6 is most likely. But while Leonard says that these three are equally spaced, this other person says that her credence of getting heads is substantially closer to her credence of rolling 3 than to her credence of rolling less than a 6. At most, Meacham and Weisberg say, we can preserve the reality of ordinal comparisons, saying which events are more likely than others, but not saying anything about cardinal comparisons of how *much* more likely one is than another.

Furthermore, Meacham and Weisberg argue that this view leaves us unable to compare the degrees of belief of different people. If degrees of belief are just a feature of what each person has, and each can be transformed in a different way, then there is no way to compare a degree of belief one person has with a degree of belief of another person.

However, I will argue that by properly understanding the analogy between the measurement of degree of belief and the measurement of quantities like distance, temperature, and so on, we can recover interpersonal comparisons of degree of belief. Furthermore, we can recover various notions of how *much* more confident one is of one proposition than another, though I think there are some conceptual worries about the extent to which this notion really is significant. While none of this establishes that degree of belief is *in fact* prior to preference, it is just meant to show that there are no significant problems on this front from assuming that it is.

4.1.3 Measurement theory

The aim of this section is to argue that degree of belief is much like temperature. Donald says that water freezes at 32 degrees and that room temperature is 68 degrees, while Justin says that water freezes at 0 degrees and that room temperature is 20 degrees. If we interpret them charitably, and are familiar with both the Fahrenheit and Celsius scales of temperature, we take this to be a difference in the numbers they use, but not a difference in the strongly real temperature underlying the two numerical descriptions. As I will argue toward the end, there is actually a similar ambiguity to the notion of "how much hotter" as there is in "how much more confident". But before I get there, I will use distance as an example to show how the numerical representation of real parts of the world works.

The central ideas were set out most systematically by Krantz et al. (1971), and my presentation follows much of their work. The concepts of length, mass, temperature, and so on aren't literally numerical. Rather, in each case, there are some real physical relations that happen to obey various laws, and since those laws have the same structure as some arithmetical laws, we can choose to represent those physical relations with numbers. Some features of this representation correspond to the real physical laws, while others are just conventional choices made to simplify our calculations. The same is true for degree of belief, and the challenge is to figure out which aspects of the representation are to be taken literally and which are to be taken as conventions. I will outline the situation for distance, then discuss some additional features that show up in the discussion of temperature, and then draw analogies for degree of belief.

4.1.3.1 Distance

For distance, we assume there are two fundamental, real relations. The first is a comparative notion that lets us compare pairs of locations. If A and B are two locations and C and D are two locations, we imagine rigid rods stretching between those locations, and think about what would happen if we were to line them up parallel to each other, starting from the same point. If the rod that started stretching from A to B

would stick out past the end of the rod that started stretching from C to D, then we say that $AB \succ CD$. Some amount of fictionalization and idealization is needed for this—there is no rod stretching from Houston to Dallas, and there is no rod stretching from Los Angeles to New York, and even if there were, no one could move them to line them up. But what is important is that there is *some* feature of the world that means that if these rods *did* exist, and someone *could* move them, the one from Los Angeles to New York would stick out past the one from Houston to Dallas. Maybe you'd need to somehow move the Rocky Mountains and relocate the population of the central United States to move these rods without breaking them. But the claim is that something like this is part of what we really mean when we say that the distance from Los Angeles to New York is 2470 miles while the distance from Houston to Dallas is only 239 miles.

The numerical representations of these distances need to correspond to this real relation. We will want to represent each pair of locations with a number in a way that $AB \succ CD$ if and only if the number corresponding to AB is greater than the number corresponding with CD. We will also eventually impose some other constraints on this representation, but in order for this sort of representation to work at all, we need the world to govern the behavior of these hypothetical rigid rods in particular ways.

The first thing that is needed is that, since no number is greater than itself, it needs to be the case that if you start with two copies of a rigid rod, there is no way to move them so that one of them sticks out past the other. This is basically what we mean by "rigid", but it is important that there really be some feature of space itself (or at least, the part of space we are considering) that supports this feature. If there are strange wormholes or ways of moving objects relativistically that deform them in this way, then our notion of distance will break down if we try to apply it in those contexts.

The second thing that is needed is that if $AB \succ CD$ and $CD \succ EF$, then $AB \succ EF$. That is, if there were three of these rigid rods and the first could be moved to stick out past the second, and the second could be moved to stick out past the third, then the first could be moved to stick out past the third. In combination with the first rule this means that no matter how you move two rods, whichever one sticks out past the end of the other always will, no matter how you line them up. This seems quite natural, but it is an empirical feature of space that guarantees this, and not mathematics itself.

One other important thing to note—since we are imagining such hypothetical rods connecting *any* pair of points in space, there is also a hypothetical "rod" AA from a point to itself, which we might think of as empty. It is important that for any two distinct points A and B, AA is not longer than BB, but AB is longer than either AA or BB.

Beyond the comparison of pairs of locations, the second real feature of the world that is needed to fix our numerical representation of distance is the idea of what happens if two of these hypothetical rigid rods are laid end-to-end in a straight line. The resulting "concatenation" of rods should behave just like one of these rods and can thus be compared to individual rods. If AB and CD are two such rods, then we will represent this concatenation by $AB \circ CD$. Again, these rods don't really exist, and it's even harder to imagine lining a pair of them up in a row and making the result rigid than it is to imagine just comparing two of them. But I claim that something about this idea is that we mean when we note that the distance from Houston to Austin of 152 miles, added to the distance from Dallas to San Antonio of 248 miles, is less than the distance from New York to Chicago of 737 miles.

We want our numerical representation to be such that $AB \circ CD$ is represented by the number that is the sum of the numbers representing AB and CD. In order for this to work, there are several more features of these rods that the world needs to ensure. First, since $x + y = y + x$, it needs to be the case that sticking two rods together in one order and sticking them together in the other order give rods of the same length. Second, it needs to be the case that if $AB \succ CD$ and $EF \succ GH$, then $AB \circ EF \succ CD \circ GH$. That is, there is no way to put together two shorter rods and get something longer than what you get if you put together two longer rods. Although these features of the physical world are so fundamental to our understanding of the world that they *feel* like mathematical truths, they are in fact substantive assumptions. In fact, much of our intuitive understanding of arithmetic probably derives from our physical experience with distances and other real-world concepts that behave in these analogous ways.

Once we have all of these features, we can begin to assign numbers to distances. For the trivial "rod" AA, we see that sticking it to the end of any other rod doesn't change the length. So if concatenation is represented by addition, then this trivial rod must be represented by the number 0. If we choose some other specific rod and call it 1, then we can begin to approximate other distances by seeing how many times this rod can go into it. Something like this is behind the idea of using one's foot to measure distances in feet. But the assumptions we have made so far don't quite guarantee that we can get precise numbers to correspond to all distances.

One quickly notes that whatever rod one uses, there are distances that don't exactly line up with any number of copies of that rod. If you lie down on the ground and mark the length from your head to your toes, it is quite unlikely that you can fit exactly 5 feet or exactly 6 feet in that length. Most likely there will be some space left over. We are used to dealing with this by choosing some shorter rod (perhaps the first joint of one's thumb, which one can call an "inch"), using it to measure the leftover, and seeing how many times this rod goes into the first one. To be sure that this technique can always work to get more and more precise numerical approximations of distance, we need it to be the case that for every rod, there is another rod that goes into it at least twice. That is, for every AB, there is some CD with $AB \succ CD \circ CD$. This question of whether space is actually arbitrarily divisible in this way has been a philosophical question since antiquity and lies at the heart of paradoxes from Zeno to Kant. But for most purposes, we can at least treat distance as if this sort of divisibility obtains.

The other fundamental assumption that is needed for this method to work is that for any measuring rod we choose (other than the trivial rod AA), and for any distance we want to measure, it has to be possible to exhaust this distance by some finite number of copies of the measuring rod. No distance can actually be infinitely far and no non-empty rod can actually be infinitely short. This is known as the Archimedean principle of distance, and again is part of our basic understanding of the world derived from experience. If you are standing in Los Angeles, it can *feel* like putting one foot in front of the other gets you no closer to New York. But if three feet are as long as a yardstick, and 22 yardsticks are as long as a surveyor's chain, and 80 chains make a mile, and there are 2470 miles from Los Angeles to New York, then there is some number of feet that gets to New York. There is even some number of millimeters to the edge of the galaxy! We discussed the Archimedean principle that Ramsey assumed for utilities in his representation theorem in Section 2.2.5—it is not a coincidence that these two representations of different systems with the real numbers use similar assumptions.

Assuming that distance actually has all of these features, so that moving rods around doesn't change their ordering, and so that rods can be concatenated in appropriate ways, and every distance can be subdivided by every other, we can represent distances with numbers.

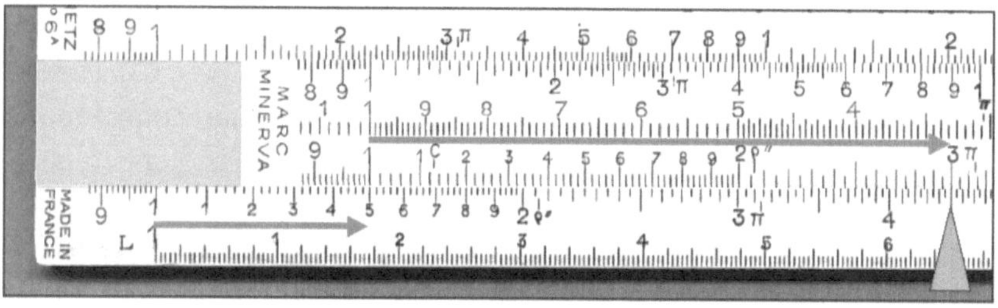

Figure 4.1 A slide rule indicating concatenation by multiplication. 1.5·3 = 4.5. (Image CC by Jean-Jacques MILAN at Wikimedia)

If we choose any unit to represent with the number 1, then all other distances can be represented in a unique way if we want concatenation to be represented by addition and greater distance by a greater number. The choice of unit is arbitrary, and once that choice is made, the rest of the representation is fixed. There is no sense in which the number 2470 is itself part of the distance from New York to Los Angeles—the number is a pure convention, and someone measuring in kilometers would represent the same distance as 3975.

While most modern people are at least implicitly familiar with this much of the conventionality of the numerical representation of distance, there is a bit more that can be abstracted away from the numbers. We made an arbitrary choice to represent concatenation of rods by *addition* of numbers, rather than some other operation. A slide rule makes use of this arbitrariness to help us calculate multiplication or other functions. Where the ordinary rule at the bottom of the slide rule in Figure 4.1 puts the numbers 0, 3, 6, etc. at equal intervals, the slide rule just above it puts the numbers 1, 2, 4, etc. at equal intervals, with the other numbers (and fractions of them) at appropriate places in between. With an ordinary ruler, sliding a length indicated by 4.9 to the end of a length indicated by 1.8 will yield a combined length indicated by their sum of 6.7. But with a slide rule, sliding a length indicated by 3 to the end of a length indicated by 1.5 will yield a combined length indicated by their *product* of 4.5. Sophisticated slide rules sometimes have multiple types of indication of distances, so that other operations than addition or multiplication can also be indicated by concatenation.

What is essential to the types of mathematical operations that can be indicated by concatenation is that they obey the properties of the actual physical concatenation operation. Concatenating a pair of lengths in either order yields the same total length, so the operation we represent concatenation with must also obey this "commutative" law. Concatenating a pair of longer distances yields a longer distance than concatenating a pair of shorter distances, so the operation we represent concatenation with must be increasing in both of its inputs. Concatenating any distance with the trivial distance from a point to itself must yield the original distance, so there must be some number that serves this null purpose in the operation we represent concatenation with. For addition this number is 0, but for multiplication, this number is 1.

If we want to make a slide rule where concatenation represents multiplication, we must mark the beginning of the ruler with 1 and then choose a particular distance, perhaps an inch, to mark with any greater number, such as 10. Points at every inch along the ruler are then marked 100, 1000, etc. To mark other numbers, we need to see how many times they multiply into any of these distances. Since $2^3 = 2 \cdot 2 \cdot 2 = 8$ and $2^4 = 2 \cdot 2 \cdot 2 \cdot 2 = 16$, we see that the number 2 must be marked at a distance where three copies is less than the distance marked 10, but 4 copies is more, so it must be somewhere between 1/4 and 1/3 of an inch. Since $2^6 = 64$ and $2^7 = 128$, we see that 6 copies of this distance must be less than 2 inches and 7 copies must be more, so it must be somewhere between 2/7 and 1/3 of an inch. By comparing further powers of 2 with other powers of 10, we can get better approximations of where exactly 2 should be marked on this slide rule, and similarly for each other number.

If we had grown up with only slide rules for measuring distance, we might naturally think of distance concatenation in terms of multiplication. This might have only been convenient if our system for writing numbers made multiplication as easy to calculate as addition actually is. Such a number system would require us to deal with astronomical numbers when indicating the distance from New York to Los Angeles or perhaps we would have had a greater variety of units than just inches, feet, and miles (or meters and kilometers). But had we represented distances in this way, the actual fact that we discuss by saying that New York to Chicago is about three times as far as Dallas to San Antonio would instead by discussed by saying that New York to Chicago is about the *third power* of Dallas to San Antonio.

4.1.3.2 Temperature

The situation for the representation of temperature has some similarities and some differences to the situation for the representation of distance.

We are familiar with the idea that the degree Fahrenheit and the degree Celsius are two arbitrary choices of unit on the temperature scale, just as the mile and the kilometer are for distance. But with distances, there is a special distance (namely that from a point to itself) that we represent with the number 0 on every scale, as long as we measure distance additively. (On a slide rule, that distance is marked with the number 1.) Our familiar everyday temperature scales instead make different conventional choices of what temperature is marked as 0.

Just as comparison of rods and concatenation of rods are the real (or at least, potential, in the case of long rods that no one will ever make) features of the world that govern the numerical representation of distances, there are some real features of the world that govern the numerical representation of temperature. For any two physical systems at equilibrium, if these two systems were to be put in contact with each other, energy would flow from one system to the other, but not the other direction.[3] If energy would move from A to B, then we say that $A \succ B$. As a matter of empirical fact, two copies of the same system would not transfer energy in either direction, and if energy would transfer from A to B and would transfer from B to C, then if A and C were in contact, then energy would transfer from A to C.[4]

The feature of the world that corresponds to the unit of temperature is a bit less familiar, but is approximately summarized by Newton's Law of Cooling. This law states that for a fixed pair of substances (whether water and copper, cast iron and onions, or a human body and the air), the rate at which energy flows from the warmer to the cooler is proportional to the temperature difference. A human body at 102 degrees in a room that is 70 degrees will chill just as quickly as a human body at 98 degrees in a room that is 66 degrees, which is why someone with a fever will feel chilly. A copper pot at 200 degrees will heat 70-degree water at the same rate that a copper pot at 300 degrees will heat 170-degree water. Coffee at 130 degrees in a 70-degree room will cool off twice as fast as coffee

[3]From a statistical mechanical perspective, this is because the total energy of the two systems is conserved, there are only a finite number of ways for energy to be distributed among all the microscopic units of the two systems, and most configurations of the microscopic units with a given total energy add up to a relatively equal distribution of energy between the two systems, which is generally not the distribution the two systems originally had.

[4]These features of the physical world basically correspond to the idea that systems always tend towards an equilibrium, rather than moving in a perpetual cycle.

at 110 degrees in an 80-degree room.[5] If one can measure the rate of transfer of energy between a pair of substances, then one can use this to mark off temperatures separated by the same distance according to this law. Whatever reference pair was used to define this difference then gets marked as a degree. If one is just measuring heating and cooling, the choice of zero is arbitrary, and the size of the degree sets the scale of energy flow for any given pair of substances.

However, temperature governs not just heat flow, but also the expansion of gases, and the efficiency of engines that turn heat into work or vice versa (whether through the use of expanding steam, exploding gasoline, expansion and compression of freon, or conversion of solar energy into electricity). Through the work of the French engineer Sadi Carnot, and later work of William Thomson (granted the title of "Lord Kelvin", by Queen Victoria), it became clear that the expansion of a gas and the efficiency of an engine or heat pump depend not on the *difference* of two temperatures, but rather on their *ratio*, when expressed in a scale with a suitable definition of 0 that we now call "absolute zero", at about −273 degrees C or −460 degrees F. As a result, there are *two* notions of the relevant temperature difference. For the purposes of figuring out how quickly energy transfers between objects, −270 C is closer to 0 C than 1000 C is—energy flows nearly four times as quickly from a source at 1000 C as it does to a sink at −270 C. But for purposes of figuring out how efficiently an engine runs, 0 C is over a fifth of the way to 1000 C, while −270 C is only about a thirtieth of the way to 0 C. Cooling the exhaust vent/intake valve of an engine can improve efficiency more than increasing the temperature of the heat source.

For some purposes in thermodynamics, it turns out that it's more useful not to represent temperature even on the Kelvin scale with absolute zero, but in terms of its *reciprocal*, sometimes called "coldness". That is, while the melting point of ordinary ice is at a temperature of about 273 Kelvin, we can say that the coldness of this state is about .00366 K^{-1}. Liquid nitrogen has a temperature of about 77 Kelvin, and thus

[5] A complexity of temperature difference that doesn't exist for distance is the dependence on the substances. The rate of energy transfer from water to human flesh is much faster than that from air, which is why (all temperatures in Fahrenheit) an 80-degree bath can feel cooler than an 80-degree room (because you're losing energy to it faster) and a 120-degree hot tub can be deadly while a 120-degree dry sauna is barely getting started (because water transfers energy to you so much faster than air). With metal it's even faster, which is why metal so often feels more extremely cold or hot to the touch than the water it contains at the same temperature.

a coldness of about .013 K^{-1}. Liquid helium reaches a temperature of about 4 Kelvin and thus a coldness of about .25 K^{-1}. When describing progress in low-temperature physics, it might seem surprising to hear just how much harder it is to get from 4 K to 1 K and then to .01 K and .000001 K. But in terms of coldness, this is the progress from .25 K^{-1} to 1 K^{-1} to 100 K^{-1} to 1,000,000 K^{-1}. This representation helps make it clear why the third law of thermodynamics says it is impossible to bring a system to absolute zero—this would involve bringing its coldness to infinity![6]

The fact that temperature and coldness are equally good numerical representations of the same subject matter and the fact that temperature differences and temperature ratios both enter into important physical laws are the two important analogies I want for degree of belief. There is no disputing that temperature is in some sense a real feature of physical systems.[7] This feature is appropriately described by fairly precise real numbers. However, the choice of Fahrenheit or Celsius is an arbitrary convention. For some purposes, we can do better by using a scale with absolute zero, like the Kelvin scale. But for those purposes, there is still an arbitrary convention of using temperature rather than coldness to describe the system. The information content is the same, though they emphasize different sorts of difference.

4.1.3.3 Degree of belief

Degree of belief has many of these same features. So far, when I have represented degrees of belief numerically, it has been with numbers that

[6]From statistical mechanics, it becomes clear that in a system of interacting particles with a given total amount of energy, there is an equilibrium distribution of what fraction of the particles will have a given energy level. It can be calculated that there will be some constant β such that the number of particles at energy level E will be proportional to $e^{-E\beta}$, the Boltzmann distribution. If β is larger, then many more particles are in low-energy states than high-energy states. As β gets smaller, more particles enter higher energy states. When separate systems individually in equilibrium are brought together, energy will flow from the system with lower β to the system with higher β, which shows that β has the opposite ordering of the characteristic comparison of temperature. It is an empirical fact that β times temperature equals a constant, which is known as Boltzmann's constant. Thus, β is a representation of the coldness, which falls more directly out of statistical mechanics than temperature itself. For more details, see Nash (1974).

[7]There are some limitations in that temperature only strictly applies to systems that are internally in equilibrium, and that systems need to be large enough to have statistical distributions in order to be characterized by temperature. But ordinary macroscopic systems usually have these features.

satisfy the probability axioms. As discussed in Section 2.1, these are the numbers that emerge when one describes the bet a person would consider fair in terms of the price paid for a ticket that pays $1 if the relevant proposition is true. If a race has three horses, A, B, and C, and one is equally confident that each would win, then one shouldn't be willing to pay more than about 33 cents for a bet that pays a dollar if the given horse wins (because otherwise one would be willing to buy tickets on all horses for more than the possible winning) and one shouldn't be willing to sell such a bet for less than about 33 cents (because otherwise one would be willing to sell tickets on all three horses for less than the amount one would foreseeably have to pay out).

However, if one actually goes to a racetrack to place bets on a horse, one will usually find these prices listed quite differently. I am told that at racetracks in continental Europe, one will find bets listed by fixing the *purchase price* of a ticket at 1 Euro, with the *payout* as the representation of the odds. If the probability of horse A winning is $1/6$, the probability of horse B winning is $1/12$, and the probability of horse C winning is $3/4$, then a European racetrack would represent the odds as 6.00, 12.00, and 1.33, respectively. However, at a British racetrack, the bets will be listed with odds indicating the *ratio* of the respective risk of the bookie and the bettor—if a ticket can pay 6.00 on a price of 1.00, then the bookie risks losing 5 units while the bettor risks losing 1 unit. Thus, at a British racetrack, the odds for horses A, B, and C would be represented as $5:1$, $11:1$, and $1:3$, respectively. American racetracks represent these even more strangely. Bets on events with probability less than $1/2$ are represented with positive numbers, while those with probability greater than $1/2$ are represented with negative numbers. Positive numbers represent the possible *profit* on a bet that costs $100, while negative numbers represent the *cost* of a bet that yields a possible profit of $100. Thus, for horses A, B, and C, an American racetrack would list the odds at $+500$, $+1100$, and -300, respectively, since a bet of $100 on horses A or B could yield winnings of $600 or $1200 (and thus profit of $500 or $1100), and a bet of $300 on horse C could yield winnings of $400 (and thus profit of $100).

These different odds listed at different racetracks form a real-life Leonard and Maurice situation. They are representing the same underlying degrees of belief on different numerical scales. The significant question is what features the underlying degrees of belief have that are being equally well represented by the three numerical scales. Distance must be stable under moving rigid rods in a way that preserves various features of comparison and concatenation, in order to be properly represented by our ordinary additive scale, but then it is also equally well represented by the multiplicative scale on a slide rule. Temperature must be related in systematic ways to the transfer of energy and efficiency of engines in order to make sense of the ordinary scales, but it can then also be represented by coldness. To address Meacham and Weisberg's concern about interpersonal comparison of degrees of belief, we need to understand this underlying real feature of degrees of belief in a way that lets us represent the degrees of beliefs of different people on the same scale.

And in fact, a version of this work was already done even before the general work on measurement theory by Krantz et al. (1971). In his (1946), the physicist Richard Cox gave a series of features that degrees of belief must have in order to be represented by probabilities and he showed that any system that has these features can be represented either by probabilities, or by any monotonic numerical transformation of them like the European, British, and American odds.[8] To fully state Cox's theorem requires the notion of *conditional* degree of belief, which will be the subject of Chapter 6. But for now, it is sufficient to say that Cox required basically just that there be some systematic way that a subject's degree of belief in a proposition is related to their degree of belief in its negation, and some systematic way that their degree of belief in a conjunction A and B is related to their degree of belief in A, and their conditional degree of belief in B, given A. Given such systematic relations for a specific subject's psychology, there is always a unique way to represent their degrees of belief on a probability scale, and a unique way to represent their degrees of belief on any other numerical scale that one chooses to use. If we can justify Cox's assertion of such systematic relations, then interpersonal comparisons are not a problem. Many physicists, like Cox and Jaynes (2003), have just assumed that

[8]Technically, Cox's Theorem contains a missing step, but this has been clarified and fixed by the mathematician Jeff Paris (1995), and separately by the computer scientist Joseph Halpern (1999a,b). Importantly, for the proof to work, for every proposition that a person has a non-minimal degree of belief in, there must be another one that the person is less confident in while still being non-minimal. A version of this assumption appeared in Savage's representation theorem, as discussed in Section 2.3.4. The set of possible degrees of belief must be infinite in order to establish unique reference points at all levels of the scale. This is no more problematic for finite beings like us than the fact that our finite vocabularies allow us to express infinitely many sentences in our native language, and that for every event, someone can consider the question of whether that event will happen and *also* an independent coin will come up heads.

Cox's assertions are "common sense", so there is no need to justify them, but I think that justifying the existence of these systematic relations is where the important work is to be done.

This is how I see much of the work of the theorems from Chapters 2 and 3. The pragmatic role of degree of belief in guiding action, or the alethic role in accurately representing the world, should not be used to justify the specific use of the probability representation on a scale from 0 to 1, any more than the measurement theory of temperature justifies the use of a scale from 32 to 212 for liquid water or the measurement theory of distance justifies the use of the king's foot as a unit. Instead, these theories should be used to explain the *systematic structure* of these quantities, which establishes that these scales *can* work, though others do as well.

The representation theorem argument considered by Zynda, and Meacham and Weisberg, does not by itself justify strong realism about degree of belief. But if we have reasons for thinking that there is such a strongly real psychological state that plays this pragmatic role, then the representation theorem argument gives us a way to understand its structure that justifies its measurement by probabilities, or by any of the other types of odds scales that have been used in gambling.

If one instead takes the alethic role as the fundamental role for this psychologically real state, as in Chapter 3, then the argument must work differently. The argument attributed there to Joyce (and developed further by Pettigrew and others) requires some very specific numerical features in a measure of accuracy, and argues that if one's degrees of belief do a good job of being accurate, then they will satisfy the probability axioms. If the argument as phrased by Joyce is correct, and accuracy really is measured in his way, then it would be incorrect to represent degrees of belief by the odds scales mentioned above. Since these scales are clearly reasonable, Joyce's argument is too strong. There must be a way to weaken his assumptions, if the general strategy is right.

And in fact there is. Lindley (1982) proved a more general theorem, though he didn't give the philosophical interpretation that Joyce did. Much of the philosophical significance of this has recently been worked out in greater detail by Ajalloeian (2025). While Joyce's second set of three axioms very specifically depend on the particular numerical representation of degrees of belief, Lindley just assumes that there *are* a highest and lowest numerical degree of belief, that there *are* a decreasing function $I(x, 0)$ and an increasing function $I(x, 1)$ that measure the accuracy of degree of belief x in worlds where the proposition is true or false respectively, and that these functions are differentiable. It's clear that if these conditions are satisfied by one numerical representation of degrees of belief, they will still be satisfied if the numbers are moved around in some way that preserves the ordering of the real numbers and is itself differentiable. With these weaker assumptions, Lindley doesn't show that degrees of belief *do* satisfy the probability axioms, but merely the claim proven by Cox that the scale of degrees of belief can be *converted* in a unique way to the probability scale. In fact, Lindley is able to concretely define the rule that transforms these numbers into the probability scale, as

$$P(x) = \frac{I'(x,0)}{I'(x,0) - I'(x,1)},$$

where each I' is the derivative of the relevant inaccuracy measure with respect to the given numerical representation of degree of belief. In fact, as he mentions, it isn't even essential that the *same* $I(x,0)$ and $I(x,1)$ are used for each proposition—all that matters is that there is some such pair of functions for each proposition, and that the overall inaccuracy of the doxastic state is the sum of these inaccuracies for the individual propositions. This still does need some sort of measurement theory for the numerical measurement of accuracy, and this theory should be detailed enough to ground the ideas of additivity and differentiability. But any differentiable inaccuracy measure will be transformed by this conversion into a proper scoring rule. There is no justification needed for propriety.

4.1.3.4 Differences in degree of belief

I think there is also another way to establish a numerical scale for degrees of belief by considering Meacham and Weisberg's second question of how *much* more confident one is of one proposition than another. They claim that the sort of flexible numerical representations considered here (where each form of odds is equally good as the representation as probabilities) undermines meaningful ideas of comparing not just credences in propositions, but *differences* in degree of belief. I think this idea of comparing differences in degree of belief is perhaps more subtle than they suggest—just as with distances and temperatures, other calculations like ratios will often be more meaningful. But I think that one way of considering these differences leads to another natural use of measurement theory to produce numerical structure for degrees of belief.

As discussed by Fitelson (1999), there are actually a variety of different measures of difference in degree of belief that have been proposed by Bayesians, even within the probability representation by itself. If degree of belief x is greater than degree of belief y, and degree of belief y is greater than degree of belief z, then everyone agrees that the change from x to z is a bigger change than the change from x to y or the change from y to z. The only case mentioned by Meacham and Weisberg (Hosiasson-Lindenbaum (1940)'s influential Bayesian treatment of the ravens paradox) happens to be a case of this form. But it is harder to compare differences in degree of belief when the differences involve two completely separate pairs of degrees of belief.

For instance, in one study, 34% of 50 year old American smokers eventually developed cardiovascular disease, while 27% of nonsmokers did (Lloyd-Jones et al., 2006). In another study, 5.2% of male Polish former smokers died of lung cancer, while 1% of Polish men who had never smoked did (Brennan et al., 2006).[9] To simplify calculations, I'll treat the change in cardiovascular disease as 33.3% to 25%, and the change in lung cancer from 5% to 1%. In terms of probability, the change in cardiovascular disease was about 8%, while the change in lung cancer was only 4%, so you might say that the particular observations yielded twice as large an effect for cardiovascular disease as for lung cancer. However, smoking brought the *odds* of getting cardiovascular disease from $3:1$ to $2:1$, while it brought the odds of lung cancer from $99:1$ to $19:1$. It is also common to report changes in terms of ratios rather than differences. The ratio of probabilities (reported in medical journals as a "risk ratio") in the cardiovascular disease study is $4/3$, while the ratio of probabilities in the lung cancer study is $5/1$. And the odds ratio in the cardiovascular disease study is $3/2$ while the odds ratio in the lung cancer study is about $99/19 \approx 5.2$. On either the risk ratio or odds ratio measure, lung cancer was much more affected by smoking than cardiovascular disease.

It's not clear to me that there is a univocal answer to which difference should be treated as larger. So I think we shouldn't give in to Meacham

[9]Numbers were much higher for people who continued smoking throughout their lives, ranging from 17% in Russia to 82% in Slovakia. The rates of lung cancer in non-smokers in these Eastern European countries were apparently also higher than the rate of 0.2% often observed in the UK or US. I don't know the full context of these numbers or what explanations there might be. I just chose the examples I did because the numerical values are a bit more convenient to work with than many of the others.

TABLE 4.1 Degrees of belief for a subject who is uncertain whether the coin is 2/3 biased towards heads or 2/3 biased towards tails.

		Heads	Tails
Biased toward heads	x	$\frac{2}{3}x$	$\frac{1}{3}x$
Biased toward tails	$(1-x)$	$\frac{1}{3}(1-x)$	$\frac{2}{3}(1-x)$

and Weisberg's demand to produce a clear distinction between "greater" degree of belief and "much greater" degree of belief. My degree of belief in a particular smoker getting cardiovascular disease is greater than my degree of belief in a particular non-smoker getting cardiovascular disease. And my degree of belief in a particular smoker dying of lung cancer is greater than my degree of belief in a particular non-smoker dying of lung cancer. But I don't think there is a fact about one of these being "much greater" and the other not—there is a higher change in probability for cardiovascular disease, and a higher risk ratio for lung cancer mortality, and these are just different ways that one degree of belief can be greater than another.

Some Bayesians (Christensen, 1999, Vassend, 2018) have suggested that these different measures should be used to compare changes in different contexts. The difference in probabilities may be the more relevant numerical measure of change for purposes involving action plans while other measures might be better for other purposes. However, as I'll show, there's at least some sense in which the odds ratio (or perhaps better, the logarithm of the odds ratio) is a natural measure for the amount of evidence one has. This may help spell out a measurement theory for an evidential role for belief, as considered in Section 3.5.

Consider a situation in which one is investigating a trick coin from a magic shop. One knows in advance that some coins from the shop are biased with a 2/3 chance of landing heads, and the rest are biased with a 2/3 chance of landing tails.[10] That is, whatever initial degree of belief x one has, represented on the probability scale, for the coin being biased toward heads, one's degree of belief that it is biased toward heads and actually lands heads is 2/3 of this, and one's degree of belief that it is biased toward heads but lands tails is 1/3 of this, and conversely for bias toward tails. These degrees of belief are illustrated in Table 4.1.

[10]Though see Gelman and Nolan (2002) for an argument that such trick coins are actually impossible!

As I will argue in Chapter 6, when one observes a heads flip, one eliminates the tails possibilities, and renormalizes the resulting degrees of belief. Thus, one's new degree of belief in heads will be $\frac{(2/3)x}{(2/3)x+(1/3)(1-x)} = \frac{2x}{x+1}$. This numerical update is hard to think about in the probability representation.

But it's easier if we represent it in the British odds. Let $O = \frac{x}{1-x}$ be the ratio corresponding to the initial British odds for the coin being biased toward heads. As I will show in Section 6.2.5, the British odds after observing a heads flip will be $\frac{(2/3)x}{(1/3)(1-x)} = \frac{2x}{1-x} = 2O$. That is, every observation of heads doubles the odds, and every observation of tails halves them.

This update is even easier to work with in the "logit" representation, which is just the logarithm of the British odds. If we use the base 2 logarithm, then each observation of heads increases the value by 1 and each observation of tails decreases it by 1. The value of 0 corresponds to odds of 1, which means equal probability of bias toward heads or bias toward tails, which means probability 1/2. Thus, on the logit representation of degree of belief, we can think of degree of belief as being something like a measure of how many independent bits of evidence would be needed to bring a proposition from perfectly balanced probability to this probability.

Thus, we can imagine setting up a quantitative notion of measurement for *amount of evidence* just as we did for distance initially. Instead of comparing distances between pairs of points AB and CD, we can compare distance between propositions (P, Q) and (R, S) in this evidential sense. For distance, we have to imagine a pair of hypothetical rigid rods stretched between AB and CD, and imagine sliding them toward each other to compare which is longer. For evidence, we have to imagine a pair of hypotheses like the ones about biased coins and figure out whether the number of independent observations needed to bring it from the degree of belief one has in P to that of Q would be more or less than the number of independent observations needed to bring it from the degree of belief one has in R to that of S.

Concatenation of evidence is simple in the case of pieces of evidence that are independent conditional on each hypothesis. When pieces of evidence aren't independent, we have to imagine replacing them with ones that are independent, just as when pairs of points aren't collinear, we have to imagine straightening out the rods so that they are lined up. If these evidential analogs of the movements of rigid rods can be made sense of, then we can directly choose some unit of evidence and get a ratio scale for it in the same way that Krantz et al. (1971) gets a scale for distance. This could be another route to measurement theory for degrees of belief if one can make sense of the relevant notion of comparative evidence without already having the mathematics of probability theory. In fact, something like this discussion was given by I.J. Good and Alan Turing in their early wartime work on codebreaking, though it was't published until much later (Good, 1985). (Interestingly, Vassend (2015) gives a very different argument that the logarithm of the change in the odds ratio is the only confirmation measure that behaves well from a measurement theory perspective.)

To summarize, I think it's not clear what the best measurement theoretic foundations are for degree of belief. However, several good starts are available. In any case, Zynda's worry about alternative numerical representations is not a worry for the strong realism of degree of belief—it is a universal feature of all numerical quantities, whether psychological or physical. Meacham and Weisberg's worries about these numerical representations can be addressed if good foundations for the measurement of degrees of belief are available. Each person's degrees of belief will have a unique representation as a probability function (and also a unique representation as British odds, or European odds, or American odds, or logit, or any other standardized numerical representation). Any such standardized numerical representation will suffice to define interpersonal comparisons of degrees of belief. Comparison of differences of degree of belief works fine in the one case that is most pressing (that of saying when one piece of evidence is stronger than another for the same proposition), while in other cases there may not be a univocal notion. But there are several different standardized calculations of change in degree of belief, any of which can be applied to any of these standardized numerical representations of degree of belief, as long as we are clear on which representation we are using. And there may even be a direct measurement-theoretic foundation for quantity of evidence, that is more naturally based in the log odds ratio.

4.2 PSYCHOLOGY

Another sort of challenge to numerical degrees of belief comes from Harman (1986) and Holton (2015). The basic idea of both authors (which others have supported) is that although it might be nice for some sort of doxastic subject to have degrees of belief that obey something like the axioms of probability theory, it would be just too complicated for humas. They argue that rather that rather than these continuously graded quantities like degree of belief and utility, our fundamental doxastic and bouletic states are constituted by attitudes like belief or intention, that are either on or off. These are not mere simplifications of the graded concepts, but are in fact the more fundamental attitudes.

This is not an objection to the broader project of Bayesianism as a theory of the doxastic states of any sort of agent that has degrees of belief—they allow that there could be more complicated doxastic subjects, like artificial intelligences, or groups like the scientific community as a whole, or possibly aliens or angels, that have these finely graded states. Their objection is targeted specifically at humans.

A full evaluation of the force of these arguments about humans would require detailed empirical investigation of the human mind that neither they nor I are in a position to carry out. My response to all of these arguments will take the form of arguing that the human mind has a lot more computational capacity than we consciously use, and this unconscious ability may well be sufficient to implement a sort of Bayesianism. Whether the human mind in fact does so is an empirical question that will take a lot more research. But even if not, I claim that a Bayesian mind would not be as different from the human mind as Harman and Holton suggest.

Harman's objection is that properly working with degrees of belief is too difficult because of the number of different propositions one must work with. As I will describe in Chapter 6, the standard Bayesian picture for how one's degrees of belief change over time requires that one have degrees of belief in every conjunction of a proposition with potential evidence. If every proposition could possibly be learned, then that means that one needs to have a degree of belief in every truth-functional combination of propositions one considers. Harman points out that this leads to a "combinatorial explosion"—with 10 basic propositions, there are 1024 truth-functional combinations, while with 20 there are over a million, and with 30 there are over a billion. He argues that having this many degrees of belief would be far beyond the capacities of the human brain.

I think this claimed limitation on the human brain is quite unclear. The brain itself is quite complex and well beyond our current understanding. We don't know what limitations it has. How many names do you know? Think of all the names of your friends and family, including "friends" that are merely connections on social media. How many more names do you know of celebrities, artists, philosophers, historical figures? Add to that all the people that you may have gone to school with, whose name you would recognize if you saw it in the right context or local business owners that you'd recognize in combination with the name of their business. There are probably thousands or even tens of thousands of such names. Is there any reason to believe that your brain is anywhere close to the limit of names you could know? With billions of neurons and who knows how many connections between them, there's no particular theoretical reason that it couldn't store simply astronomical amounts of information. Furthermore, nothing about the Bayesian idea requires these degrees of belief to be stored independently of each other—perhaps the brain works in some sort of compressed way, where degrees of belief in many related propositions are stored in some unified way that doesn't require each one to have its own separate representation. Stereotypes, biases, and heuristics are among the ways that we store many degrees of belief with only a small amount of cognitive power.

While Harman thinks it a priori unlikely that humans *could* have minds powerful enough to deal with the large numbers of degrees of belief involved in Bayesian reasoning, Holton thinks that there are reasons to believe that we *don't* have degrees of belief (or "credences") of this probabilistic sort. He says, "the mental states involved here are nothing like credences. They do not typically register anything like numerical degree of probability; we cannot manipulate them; they do not obey anything like" the mathematical rules they are said to obey. He says, "It is possible that people had been entertaining and manipulating credences for millennia ... without realizing what they were doing, but it strikes me as implausible." The mathematics of probability theory is complex and difficult, and was only developed in the past few centuries. Holton suggests that therefore this can't be the sort of math that is going on in our minds.

I think this argument is based on a mistake like that of Zynda in arguing against strong realism. Both of these arguments seem to think that a strong realist view of an aspect of thought requires that we be aware of this aspect of thought. However, it's clear that there are many features of our mental life that we aren't directly aware of. Psychologists claim to have discovered many phenomena of the human mind—cognitive dissonance, confirmation bias, risk aversion, REM sleep. It seems quite plausible that people's thoughts had exemplified these phenomena for ages without realizing what they were doing.

This is true even for some quite complicated numerical phenomena. The human visual system is capable of quite complicated calculations. When trying to catch a ball thrown in a high arc, a person trying to catch the ball must predict where it will end up, which (according to Newtonian physics) involves computing a second-order differential equation or at least tracing out a parabolic arc. However, people were able to catch thrown objects long before ancient mathematicians worked out the concept of a parabola, which itself happened long before early modern physicists argued that a parabola was the trajectory of a falling object. We don't need to be aware of a mathematical feature of our mind in order for it to work mathematically. This is perhaps even clearer for features like blood pressure, blood sugar, and neuron excitation, which can be observed to affect (or perhaps even constitute) thinking in a variety of ways, and are precisely numerical, but are very difficult[11] for people to introspect.

Holton further argues that not only are we not aware of any probabilistic reasoning, but in fact we do quite badly at it in many cases when it is made explicit.

> If we are to be good subjects for the ascription of probabilistic attitudes, then we should be able to make the kinds of transitions—whether in beliefs or in behaviour—that would be expected. Yet a host of now very familiar research shows that, in many cases, we are very bad at this.

(For many examples of this research, see Kahneman (2011).) The example Holton focuses on is the following:

> The probability that a woman has breast cancer is 1%. If she has breast cancer, the probability that a mammogram will show a positive result is 80%. If a woman does not have breast cancer, the probability of a positive result is 10%. What is the probability that a woman who has a positive mammogram result has breast cancer?

As he notes, the mathematics of probability says that the probability of having breast cancer given a positive mammogram result in this case is

[11] Though not impossible—people often become aware of their "blood boiling" as blood pressure increases or the sensation of getting "hangry" from low blood sugar. I suspect that with proper feedback, people could learn to give good estimates of their blood pressure and blood sugar levels by introspection.

less than 10%, but most people will answer that the probability is quite a bit higher.

However, this example seems to me irrelevant to the question of whether the mind works probabilistically. Consider a parallel case involving a physics problem:

> The velocity of a thrown baseball is 6 m/s at an angle of 60 degrees upwards from the horizontal. The acceleration due to gravity is 9.81 m/s^2 downwards. How far will the ball travel by the time it has fallen down to the same elevation from which it was thrown?

If you're anything like me, this question will be even harder to answer correctly than the question Holton considers. However, if we were actually to be in the situation described, rather than reasoning about it in the abstract, we would probably do pretty well at trying to catch the ball. It is extremely difficult to take a verbal description of probabilities, or velocities, and translate it into our credal or visual system, and then translate the answer we receive there back into numerical form. But this doesn't seem to me to give any reason to doubt the existence of some quantitative representation of these things in the mind.

When looking at a scene of people running around a playground and throwing balls, one can see which people and objects are located closer or farther, which are taller or shorter, and which are moving faster or slower. But when asked to give numerical descriptions of the heights, distances, and speeds, one is unsure. Conversely, when looking at this scene and trying to imagine a 6-foot tall person throwing a ball at 20 miles per hour, one is unsure quite how this would compare with everything else one sees. But if one in fact saw the person throwing that ball, it wouldn't be hard to catch. Similarly, when considering the uncertainties in your daily life, you have some sense of whether it's more likely that you will get stopped by a red light, or that the store will be out of yogurt, or that you will encounter your neighbor on the way out of the house. When asked to imagine a 20% chance of rain, you'll be unsure which of these events are more or less likely than it. But when presented with clouds with a particular distinctive look (that happens to be the look of 20% chance of rain in the next hour), you might do better.

The response I give here invites another challenge that is different from the one Holton makes. He raises worries about bad explicit reasoning with probabilities, which I think is no worry at all for there being a mental fact about the strength of degree of belief that is probabilistic.

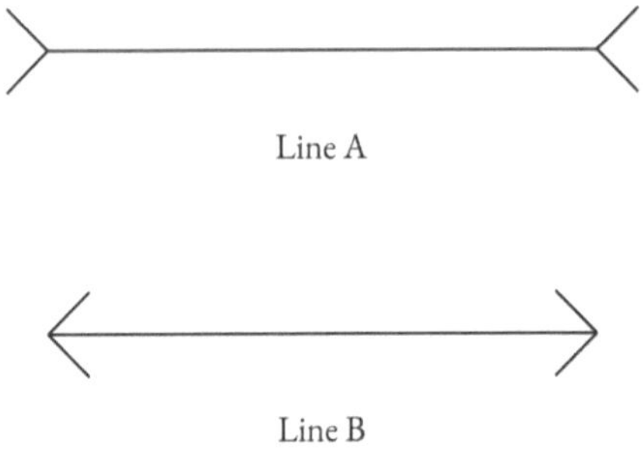

Figure 4.2 The Müller-Lyer illusion. Lines A and B are precisely the same length.

But there are also some well-known biases and errors that people make even when reasoning natively with degree of belief. People tend to feel a greater risk of crash when flying in a commercial airplane than they do when driving. And yet, as a matter of fact, there is a far greater risk of a crash when driving than flying.

This I think does point to a real inaccuracy in the degree of belief we have in various propositions. But I think this is no more problematic than various visual illusions we are vulnerable to. Just because we are bad at turning our visual impressions of trees or buildings into correct verbal expressions of the height of those trees and buildings doesn't mean that our visual impressions fail to have quantitative structure that represents those heights in ways that behave like heights. Furthermore, in many cases, the visual and cognitive illusions we fall prey to are approximations that our mind uses to quickly and conveniently represent the world in ways that are usually fairly accurate, but have some systematic flaws. Our visual system has well-known illusions like the Müller-Lyer illusion, as demonstrated in Figure 4.2.

But as it turns out, in ordinary observations of three-dimensional scenes, lines bookended with inward-pointing markings *do* tend to be images of shorter objects than lines bookended with outward-pointing markings (Howe and Purves, 2004). Shapes like Line A appear in the back corners of rectangular rooms, while shapes like Line B appear on the front corners of rectangular objects, so shapes like Line A tend to be parts of longer edges than shapes like Line B, when they take up similar lengths of our visual field. This is a useful heuristic that usually improves our estimate of the lengths of three-dimensional objects, but tends to lead us astray when looking at two-dimensional drawings. Similarly, the various means by which we come to have greater or lesser degrees of belief may do well in ordinary reasoning, even if they do badly in certain parts of the modern world with its novel media and information environments. (For more on these visual and cognitive illusions, see Chapter 8.)

A lot of mathematical sophistication is needed for doing detailed reasoning *about* degrees of belief. But one doesn't need this mathematical sophistication to *have* them. One has a height, and a neural structure, and a whole cognitive structure of beliefs and desires with their logical relations to one another without having to be aware of it. Just as Molière's "Bourgeois Gentleman" was speaking prose his whole life without having realized it, I claim it is quite possible that we have been reasoning probabilistically with degrees of belief our whole lives without realizing it. Our minds are in many ways quite a bit more sophisticated than we are explicitly aware of, and this may be another such way.

4.3 IMPRECISION

One last challenge to the claim that we really have degrees of belief that are probabilistic is the idea that it is implausible that our degrees of belief are quite so precise. There are a continuum of real numbers between 0 and 1, and it might seem odd that our finite minds can occupy those states so precisely. Many theorists have been motivated by the idea that although degrees of belief might be *like* probabilities in some way, they can't be fine-grained enough to actually *be* probabilities, even in the attenuated sense suggested here.

Much recent philosophical literature on this idea suggests thinking of degrees of belief not as individual real numbers, but instead as *sets* (usually intervals) of real numbers. (See, for instance Sturgeon (2008) and the reply by White (2009); Hájek and Smithson (2011); Elga (2010) and the reply by Moss (2015a); and Schoenfield (2016).) Some of this literature recapitulates an earlier debate. Jeffrey suggests (Levi, 1974, Jeffrey, 1984), and (Levi, 1985) that it may be hard to pin down a precise number that represents a particular person's degrees of belief, while Levi suggests that degrees of belief may even behave in more "indeterminate" ways that don't correspond to *any* number. There are several different motivations that may lead us to study theories of degrees of belief that don't work like precise real numbers. (Many such theories are described by Walley (1991).) But I will address one prominent style of arguing for this.

Miriam Schoenfield gives the following case:

> DETECTIVE CONFUSO
>
> You are a confused detective trying to figure out whether Smith or Jones committed the crime. You have an enormous body of evidence that you need to evaluate. Here is some of it: You know that 68 out of the 103 eyewitnesses claim that Smith did it but Jones' footprints were found at the crime scene. Smith has an alibi, and Jones doesn't. But Jones has a clear record while Smith has committed crimes in the past. The gun that killed the victim belonged to Smith. But the lie detector, which is accurate 71% percent [sic] of the time, suggests that Jones did it. After you have gotten all of this evidence, you have no idea who committed the crime. You are no more confident that Jones committed the crime than that Smith committed the crime, nor are you more confident that Smith committed the crime than that Jones committed the crime. (Schoenfield, 2012)

The important point she raises about this case is that it exhibits "insensitivity to mild sweetening". If you were *equally* confident that Smith committed the crime as Jones, then if a single extra eyewitness testified that Smith did it, you would now lean toward Smith. However, from the construction of the case, it seems clear that you would be just as confused and exhibit the same indecision between the two even with the additional eyewitness.

However, I think that when we compare this sort of case to other sorts of features that we unhesitatingly report with precise real numbers, we can get a better understanding of how to think of them. Consider the height of a building. In 2009, the highest point in any building in the world was the tip of the Willis Tower (formerly Sears Tower). The Willis Tower is 527 m (1729 ft) high. However, if you look closely, the tip consists of two antennas that extend from the top, which had been replaced in 2000. If you only count the top of the architectural space, it is only 442 m (1450 ft), while Taipei 101 is 509 m (1671 ft) high. Looking more closely still, you'll note that the top of Taipei 101 is an architectural spire that is not itself occupied. Its highest occupied floor is only at 438 m (1437 ft), while the Shanghai World Financial Center has its top floor at 474 m (1555 ft). (Willis Tower's is only at 412 m (1354 ft).) Which building was actually the tallest in 2009? There's a case to be made for each of them. Increasing or decreasing the height of any of these buildings by a few meters wouldn't change this confusion and indecision. (These were all surpassed by the Burj Khalifa in 2009, which is currently the tallest on all measures—tip and architectural roof at 828 m (2717 ft) and top floor at 584 m (1918 ft).)

This is one way I claim that indeterminacy about degrees of belief could result. Just as it is vague which part of a physical object is part of a building for the purpose of determining its height, it could well be vague which aspect of her mental life is an subject's degree of belief in a proposition.

Even once we have determined which features count for determining the height of a building, there is vagueness at a more fine-grained level. The steel, glass, concrete, paint, etc. that make up a building have surfaces that are constantly weathering and changing their structure at a molecular level. And the molecules themselves have indeterminate boundaries in their electron clouds (and at a much smaller scale, in their nuclei as well). The wavefunction may be defined in a perfectly determinate way, and features like heights can still be indeterminate.

However, on another level, the height of a tall building is variable over the course of a day, due to slight changes in size of metal spires as they heat and cool, and due to slight swaying in the wind (particularly when large storms come through). Thus, any report of the height of a building, if it is particularly precise, must be a report on the height at a particular moment in time or the average height over a standard time period.

Similarly, degrees of belief are presumably constantly fluctuating in real people. When considering the Detective Confuso case, I think it is natural to think that one goes back and forth as to which suspect is more likely to be guilty. This "going back and forth" might naturally be described by saying that one has degrees of belief at each moment, but as one brings different aspects of the evidence to attention, one's degrees of belief fluctuate up and down. Furthermore, shifts in blood sugar level, temperature of the room, blood pressure, and environmental factors, can often lead to shifts in attitude (regardless of whether or not these shifts are rational). Some of the fluctuation is due to attention to evidence and other parts of it are due to factors that don't have any subjective presence. If this fluctuation is fast enough and large enough, it can give rise to cases that feel like the Detective Confuso case. At one phase in reasoning, one's degrees of belief in each suspect being guilty fluctuate up and down past each other. After gaining some new evidence, it may be that one of them fluctuates around a slightly higher level than before, but they still pass each other quite regularly.

This fluctuation of degrees of belief may be merely slightly variable, like the height of a building, or much more variable, like the depth of the water in an ocean with waves and tides, or the height of the trees in a forest as they sway in the breeze, or the height of the grasses in a field as they tremble in a storm. It may be slightly misleading to represent the height of a tree in centimeters or the depth of the grasses in a field in millimeters. But there is in fact some fraction of a second in which one truly can say that the depth of the grass was exactly 543 mm (at least, up to the resolution of the surface of the solid molecules). Similarly, I claim that in many of these cases, it may well be true that at each moment, one has specific degrees of belief that can be described with a precise real number, but since one's degree of belief is shifting so unsteadily, just a few seconds of thinking about the example can make it seem like there is no fact of the matter about which proposition one is more confident in.

Degrees of belief may have some sort of vagueness of each of these sorts—they may be indeterminately constituted by components that themselves have indeterminacy (like whether the height of a building includes the spire), and they may have some sort of stable or unstable variance over short time periods (like the height of the grasses on a windy prairie). It may still be useful to describe them numerically in fairly precise ways, whether it is the way we say the Burj Khalifa is 828 m high or only the way surfers might say that (as I write this), the wave forecast for Santa Cruz is 4–6 ft high today and 6–8 ft high two days from now. Setting up numerical degrees of belief by means of a representation theorem from measurement theory doesn't guarantee that they are more precise than the surf conditions at a time (which themselves are measured by meter sticks that satisfy the assumptions of their own measurement theoretic representation theorem), but it is surely open to empirical verification whether they are more like that or more like the heights of buildings.

RESEARCH QUESTIONS

1. Is there a measurement theory for degrees of belief and desire that can underlie the strong realist notions or must one be a weak realist and construct them out of preferences?

2. Can one give a measurement theory for accuracy that will underlie the scoring rules used in Chapter 3?

3. Can a measurement-theoretic foundation for degrees of confirmation pick out the odds ratio, or are there different purposes served by different measures?

4. Can a weak realist about degree of belief give a good measurement theory for degree of confirmation?

5. Does cutting-edge empirical psychology support or undercut the idea that the unconscious mind works probabilistically? (For an example of work that begins to address this question, see Goodman et al. (2015).)

6. Are there cases where the idea of precise numerical degrees of belief for actual humans breaks down in ways that can't be captured by the fluctuation picture?

CHAPTER 5

Zero isn't nothing

IN THE PREVIOUS chapter I considered the general question of what "probabilism" must mean. I argued that the numbers involved that satisfy the axioms of probability theory must merely be a representation of some more fundamental properties and relations that degree of belief itself has. Probabilities are just one way to represent them, and other representations would work equally well (particularly the odds representation). While some argued that there couldn't really be numbers in the head, and that humans couldn't work with a representation that complex, or that degrees of belief must sometimes be imprecise, I argued that none of these were essential problems for Bayesianism as a theory of human minds (or, indeed, many other doxastic subjects).

However, some have argued in the opposite direction—that in fact, the mathematical representation must be even finer than the real numbers. In this chapter, I consider some features of the numerical representation of degrees of belief in closer detail to address some of these issues.

5.1 REGULARITY

One central feature of probability theory is that if a proposition corresponds to the empty set (meaning that there are no doxastic possibilities on which it is true), then the probability of that proposition is 0. This means that if one is certain that a proposition is false, then one should have the lowest possible degree of belief in it, which is represented by the number 0. Many authors have also endorsed the converse—if one's degree of belief in a proposition is 0, then it should be empty. Perhaps more perspicuously, if there is some doxastic possibility on which a proposition

 DOI: 10.4324/9781315626918-5

is true, then one's degree of belief in it should be a positive number, not 0. Principles of this sort have been known by the name of "Regularity" or "strict coherence".

The central motivation for this principle is a sort of faithfulness in the representation. If a and b are two propositions, and if we let "$a \succ b$" mean "the subject is more confident in a than in b", and if we let $P(x)$ be the numerical representation of the subject's degree of belief in x, then we might ask that $a \succ b$ if and *only* if $P(a) > P(b)$. It is natural to assume that if A is something the subject thinks might be true, while B is something that she doesn't, then she is more confident in a than in b. But in this case, $P(b) = 0$, so we should have $P(a) > 0$. However, there is a certain kind of case that causes problems for this principle, if the values of P are assumed to be standard real numbers. Thus, I will argue that the available options are to either modify the principle slightly (I will recommend that we accept that if $P(a) > P(b)$, then $a \succ b$, but not the converse) or to find some structure other than the real numbers to form the mathematical representation of degrees of belief.

The primary sort of case that causes problems for regularity is one in which there are an infinite number of alternate possibilities that the subject is considering. Straightforward examples arise in scientific practice all the time. A scientist is trying to figure out the value of some parameter (maybe the speed of light, maybe the electrical conductivity of copper, maybe the mass of the earth, or maybe something else). Her information so far is sufficient to rule out some possibilities completely (for instance she might be certain that the speed of light is greater than $2.9 \cdot 10^8$ meters per second but less than $3.0 \cdot 10^8$ meters per second), but there is still some range over which she is uncertain. She might be quite confident that the value is likely to be near the middle of that range, but there is still some range of values that she is unable to fully rule out. She finds every value in that range epistemically possible as the value of the parameter.

5.1.1 The cardinality argument

However, it is not consistent with the probability axioms that every value in that range has a positive probability. For those who are familiar with the set-theoretic notion of cardinality, an argument is easy to run. There are uncountably many possible values for the parameter in any interval. For each value that has a positive probability, there is some positive integer n such that the probability that that value is actual is greater than $1/n$. However, since these possibilities are incompatible, there can be no more than n of them that are greater than $1/n$, for each positive integer n. That is, for every positive integer n, the number of values whose probability is greater than $1/n$ is finite. There are only countably many positive integers, so the set of all values whose probability is positive is a countable union of finite sets and is thus countable. But there were uncountably many possible values for the parameter in the interval. Since only countably many of them have positive probability, this means that most of them do not have positive probability, and thus must have probability 0, despite being possible.

Given the concern with the meaning of real-valued representation, one might object to the use of real numbers in the above argument. After all, what the scientist is uncertain about is speeds, and speeds are not themselves real numbers, though they can be represented as such. Perhaps there are actually fewer speeds than there are real numbers, so she can have a degree of belief for each of them that is represented with a positive real number.

However, it seems perfectly consistent with what we know that there really are as many speeds as there are real numbers. Many of the physical laws we think we understand are phrased in terms of differential equations that seem to assume a structure very much like the real numbers, that would have enough possible values within it to run a version of this argument. Furthermore, nothing about the example requires the world to *actually* have speeds represented by each real number—as long as the scientist finds it *doxastically possible* that any given real number could represent the speed, then she must have doxastic possibilities with probability 0. Such a doxastic possibility is one that I will call "miniscule".

5.1.2 The continuity argument

Even without using the set-theoretic concept of cardinality, or the full structure of the real numbers as representatives of speeds, we can also give a similar argument, if the scientist is assumed to have continuous credences. The sort of continuity I want to assume is that for any speeds $a < b$, there is some x strictly between a and b such that the scientist's credence that the speed is between a and x is at least half and no more than twice the scientist's credence that the speed is between x and b (x itself should be included in precisely one of these intervals). I don't assume that the speeds are real numbers, just that they are linearly ordered, that between any two speeds there are other speeds, and that there is always a way to subdivide any such interval in such a way that the scientist's credences are not too lopsided between the two parts. Given these assumptions, we can show that for any *specific* speed, the scientist's credence that the speed of light is *precisely* that value must be less than any positive amount. As long as some of these speeds are doxastically possible for the scientist, these possibilities will be minuscule.

To see this, let c be any precise speed. Let a be some speed below c, and b be some speed above c. By the continuity assumption, there must be some x_1 such that the probability that the speed is between a and x_1 and the probability that the speed is between x_1 and b are both at most $2/3$. Let S_1 be whichever interval contains c. (Since x_1 is included in exactly one of these intervals, exactly one of these intervals will include c.) Since the credence that the speed is in S_1 is at most $2/3$, and since the speed being c entails that it is in S_1, the credence that the speed is c must also be at most $2/3$.

But now, we can apply the continuity assumption again to S_1—there is a value x_2 such that no more than $2/3$ of the credence that the value is in S_1 is below x_2, and no more than $2/3$ of this credence is above x_2. Again, let S_2 be whichever of these sub-intervals contains c, and we can see that the credence that the value is in S_2 is at most $(2/3)^2$, and so the credence that the value is c is at most $(2/3)^2$. We can keep repeating this process of subdividing the interval and show that the credence that the value is c is at most $(2/3)^n$. By choosing n large enough, we can make this lower than any positive real number.

5.1.3 Infinity is everywhere

An objection that one might raise to this sort of example is to say that there just *aren't* in fact infinitely many doxastic possibilities of the speed of light for the scientist because she is a human with a finite mind (or an artificial intelligence with only finitely much storage space or a group with only finitely many members) and can thus only think about a limited range of these possibilities. However, nothing about the notion of degree of belief requires each proposition to be explicitly represented in the brain. Much of belief and confidence is dispositional.

If we are talking about an ordinary scientist, then for every rational number p/q, there is a sentence in her language expressing that the speed of light is faster than p/q meters per second. She could well have dispositions to respond to these sentences in way that imply degrees of belief for every such interval. Very likely, for ordinary people, this process breaks down at some point, but a particularly disciplined person (or a group, or artificial intelligence) could well have this whole set of dispositions.

Thinking about things this way suggests that the problem of these infinite sets of possibilities arises in all contexts, not just scientific ones. Our linguistic and mental capacities suffice to generate infinitely many thoughts. (Thoughts about numbers are just one obvious infinite family, but there are plenty of others. I can think highly of you, and I can think that you think highly of me, and I can think that you think that I think highly of you, and so on.) As long as we are ignorant about much of the world, there are going to be uncountably many doxastic possibilities for us, at least consisting of the various recombinations of these infinitely many thoughts we can express. Some of these possibilities can be ruled out by what we know. But in most scenarios, our limitations as finite subjects mean that uncountably many of these possibilities will be around at any point.

Even if we're only distinguishing them into finitely many classes (the coin landing with the heads side up, the coin landing with the tails side up, the coin landing precisely on its edge), it may well be that some of these classes consist of infinitely fewer possibilities than others, and thus our degree of belief may need to be represented by a number less than $1/n$ for any integer n. As long as degrees of belief are represented with real numbers, this means that it must be 0, even though there is still some doxastic possibility.

5.2 IMPLICATIONS

The basic idea of representation of degree of belief by real numbers is this. Each degree of belief should have some real number that represents it. Thus, if a subject's degree of belief in two propositions is represented by different numbers, then she must actually have different degrees of belief in them. We might ideally like the converse to hold as well—if a subject has different degrees of belief in two propositions, then these different degrees should be represented by different real numbers. But I have argued that this will not in general be possible, consistent with the ordering claims that we would like, and the structure of degrees of belief in infinite scenarios.

Thus, what I will require is that if $P(a) > P(b)$, then $a \succ b$. However, I will not require the converse. If $a \succ b$, then $P(a)$ *may be* greater than

$P(b)$, but it may also be equal. We can also consider a relation $a \succeq b$, meaning that the subject is *at least* as confident in a as in b (she may be more confident or equally confident, but not less). If degrees of confidence are linearly ordered (which is one of the assumptions that is necessary for any numerical representation of degrees of belief to succeed), then for any a and b, either $a \succ b$ or $b \succeq a$. Thus, the assumption that if $P(a) > P(b)$, then $a \succ b$ is equivalent to the claim that if $a \succeq b$, then $P(a) \geq P(b)$.

The ordering between degrees of belief must be very much like the ordering between the real numbers representing them. However, it may be that different degrees of belief are represented by the same number. Once we have recognized that this holds at the bottom of the scale (both impossible propositions and minuscule ones, which have degree of belief lower than any positive real number but are still not impossible), we should recognize that the same could hold true at any point in the scale. My degree of belief that the coin will land heads is adequately represented by the number 1/2. But so is my degree of belief that *either* the coin will land heads *or* the speed of light is *exactly* $2.998018383 \cdots \cdot 10^8$ meters per second, even though I recognize a possibility where that proposition would be true without the coin landing heads. My confidence in the latter claim is ever so slightly higher than my confidence that the coin will land heads, but both must be represented by the number 1/2.

This naturally raises the question whether *every* pair of degrees of belief are strictly comparable. The argument above shows that some *particular* degrees of belief that are represented by the same real number must be linearly ordered—my degree of belief in the disjunction was higher than my degree of belief in one of the disjuncts. But it's not obvious that for *every* pair of degrees of belief represented by the same real number, one must be higher than the other. In a continuous distribution like that described in Section 5.1.2, we can argue that the scientist's degree of belief in the speed of light being any particular value must be less than any degree of belief represented by a positive real number. I have argued that most of these degrees of belief must be strictly greater than the degree of belief in the empty proposition. But it's hard to say whether each of these degrees of belief must be the same, or whether some of them might be strictly greater than others, or whether some of them might be neither greater nor lesser than others.

Since degree of belief is connected in various ways to other psychological states that are also represented numerically, similar considerations must hold there.

5.2.1 Pragmatic concepts

In decision theory, we have utility and preference as well as degree of belief. If preference can be calculated from utility and degree of belief by means of the expected value formulation, then regardless of what utility is like, the same sort of issue must arise. I'd choose to bet that *either* the coin will land heads *or* the speed of light is *exactly* $2.998018383 \cdots \cdot 10^8$ meters per second; rather than betting just that the coin will land heads. The former bet has some possibility of winning where the latter bet does not. However, the expected value of either bet is half of the possible winnings. I prefer one to the other even though the expected value is the same. I can insist that whenever the expected values are *different*, one should prefer the one with the greater expected value, but allow that when the expected values are the same, one *might* be indifferent between the two—though one also might prefer one to the other. (I show how to derive a notion of preference like this from degree of belief and utility in my (2014a).)

Interestingly, this might require some slight modification of the arguments for probabilism based on decision theory. Some of those arguments require that strict preferences *always* correspond to different expected values. For instance, Jeffrey (1965) assumes an axiom known as "continuity", while Ramsey (1926) assumes an "Archimedean" principle. Either principle implies that if $A \succ B \succ C$ (where in this case, A, B, C are possible actions and $\succ$ represents strict preference between them), then there is some finite n such that $A \succ B \succ B_1 \succ B_2 \succ \cdots \succ B_{n-1} \succ C \succ B_n$, where the amount by which one prefers B_i to B_{i+1} is always exactly equal to the amount by which one prefers A to B. However, if A and B are degrees of preference that are represented by the same real number, while C is represented by a lower one, then no finite number of such steps can suffice.

Thus, one must either choose a different representation theorem (such as the one by Savage (1954)) or else take an alethic view of the fundamental defining role of degree of belief, and use something like the method I discussed in Section 3.6.1 to construct preference from it.

5.2.2 Alethic concepts

Similarly, when considering alethic value, we must be able to say that there are different degrees of belief represented by the same real number. This raises some problems for the measurement of alethic value. No specific measurement theory was given to explain the meaning of the real number representation of overall alethic value, and now we see some further problems. The argument that the degrees of belief could be represented by individual real numbers relied on the fact that inaccuracy was represented by real numbers, and that there were only two truth values, so out of any two degrees of belief, one must be more inaccurate when the proposition is true and the other must be more inaccurate when the proposition is false. If it is possible for there to be two distinct degrees of belief that are both best represented by the same real number, then it is hard to say what their inaccuracy must be like. Because the measure of inaccuracy was continuous, these distinct degrees of belief that are represented by the same real number can't have distinct numerical inaccuracies—any number greater or less than the inaccuracy of one of them would be greater or less than the inaccuracy of a degree of belief represented by a slightly higher or lower real number as well! Thus, if there are distinct degrees of belief represented by the same real number, this must mean that there are distinct degrees of inaccuracy also represented by the same real number. Whatever these degrees of inaccuracy are, much of their structure will be inherited by the degrees of belief.

5.3 ADDITIVITY

In this section, I will freely appeal to Cantor's theory of infinite cardinalities and the distinction between countable and uncountable infinities.

I have mentioned on a few occasions that non-negativity, normality, and additivity are not quite the complete axiomatization of probability. The last axiom is a generalization of additivity to infinite cases.

Recall that additivity states that the probability of a union of two disjoint propositions is the sum of their probabilities. It's straightforward to see that this automatically applies to the union of any finite collection of disjoint sets. One might naturally want to generalize this to *all* collections of disjoint sets.

However, this generalization requires some explanation of what the sum of infinitely many numbers would be. Working purely mathematically, without any attempt to connect to measurement theory for mental or physical quantities, there are several standard theories of infinite addition.

5.3.1 Mathematical summation

The most commonly discussed theory of infinite addition applies only to the sum of countable sequences of real numbers and depends in important ways on the order in which the terms are summed. This version of summation is typically represented with a capital sigma and a term that is indexed with the natural numbers, like "$\sum_{i=1}^{\infty} a_i$". This sum is defined to be the limit of the partial sums, $\lim_{n\to\infty} \sum_{i=1}^{n} a_i$. This turns out to generally depend on the order of the terms. For instance, consider the sums:

$$1 - \frac{1}{2} + \frac{1}{3} - \frac{1}{4} + \frac{1}{5} - \frac{1}{6} + \frac{1}{7} - \frac{1}{8} + \frac{1}{9} - \ldots,$$

and

$$1 - \frac{1}{2} - \frac{1}{4} + \frac{1}{3} - \frac{1}{6} - \frac{1}{8} + \frac{1}{5} - \frac{1}{10} - \frac{1}{12} + \ldots.$$

Both of these sums have all the same terms in them. But we can see that the partial sums of the second must converge to half of the value that the partial sums of the first do:

$$(1 - \frac{1}{2}) - \frac{1}{4} + (\frac{1}{3} - \frac{1}{6}) - \frac{1}{8} + (\frac{1}{5} - \frac{1}{10}) - \frac{1}{12} + \ldots.$$

The first converges to $\ln 2$, so the second must converge to $\frac{\ln 2}{2}$. In fact, since the positive terms of this sequence and the negative terms of the sequence both individually converge to 0, but have infinite sum, it turns out that for any real number x, it is possible to choose an order for the terms that sums to x. (Start by adding enough positive terms in sequence to get above x. Then sum enough negative terms to get back below x. Then sum enough positive terms to get back above x. And so on. Because the positive terms sum to infinity, and so do the negative ones, it is always possible to continue this process. But because the terms go to 0 individually, these partial sums overshoot x by less and less, and the partial sums must eventually remain close to x.) Fortunately, these problematic features of this theory of infinite summation aren't relevant for probabilities, because they require there to be both positive and negative terms, and require those terms to sum to infinity.

The next most commonly discussed theory of infinite addition is more directly relevant. This one applies to arbitrary sets of terms that are all non-negative, that don't necessarily have an order. This is sometimes notated as $\sum_{i \in I} a_i$. This concept of summation says that the sum of the set of terms is the least real number that is at least as large as every partial sum of finitely many terms. It turns out that if the sum of the set is finite, then all but countably many terms of the set must be zero. (In order for the sum to be finite, there must be only finitely many terms greater than 1 or else arbitrarily large numbers could be achieved as finite sums of these. Similarly, there must be only finitely many terms greater than 1/2, and only finitely many greater than 1/3, and so on. Since every non-zero term is greater than some $1/n$, there can only be countably many such non-zero terms.) This is the theory of summation that is most often proposed as applicable to infinite sets of probabilities.

5.3.2 Full, finite, or countable additivity

But if we require arbitrary collections of disjoint propositions to satisfy additivity, an axiom known as "full additivity", then we rule out continuous distributions like the one motivated in Section 5.1.2. As I said there, for a continuous distribution, every single specific possibility has probability 0. But under the theory of infinite summation as the least upper bound of the finite sums, the sum of any number of 0s is 0, when the set of all possibilities should have probability 1 to satisfy normalization. Since continuous distributions are relevant for much of science, full additivity is not part of the orthodox mathematical theory of probability. (Because full additivity requires that the sum of all probabilities is 1, and the sum of a set of terms is finite only if it has countably many non-zero terms, the mathematical theory of fully-additive probability functions is basically equivalent to the branch of probability known as "discrete probability", on which there are only countably many possibilities, each with non-zero probability.)

Some theorists, like de Finetti (1974) and Seidenfeld et al. (2024) respond by denying that any principle of infinite additivity should be added—the theory of probability is complete with just the finite axioms. They develop a theory of finitely-additive probability functions for infinite probability spaces. But it turns out that this theory is exceedingly mathematically complex. It requires the set-theoretic Axiom of Choice to even prove that non-trivial finitely-additive probability functions exist. With mere finite additivity, it is straightforward to prove that the probability of a union of a collection of disjoint propositions is at least as large as the sum of the probabilities. But if the collection of disjoint propositions is infinite, then the probability of the union can be strictly larger than the sum of the probabilities. Specifying the probabilities of all these infinite unions when it isn't determined by the probabilities of its parts is where the complexity comes in.

Instead, the orthodox mathematical theory of probability adopts the axiom known as "countable additivity". The union of countably infinite collections of disjoint propositions has probability equal to the sum of the probabilities of these propositions. But for uncountably infinite collections (such as the collection of all propositions that the speed of light is precisely x), the probability of the union can be greater than the sum of the probabilities.

In order to avoid the complexities of finitely-additive probability, the orthodox mathematical theory of probability doesn't claim that *every* set of possibilities is a proposition that has a probability. Instead, there is just a "σ-algebra" of propositions that have probabilities. What it means to be a σ-algebra is that it is a collection of sets such that the complement of any set in the σ-algebra is also in the σ-algebra, and the intersection or union of any countable collection of elements of the σ-algebra is in the σ-algebra. Furthermore, this σ-algebra is usually "countably-generated", meaning that there is some countable collection of elements such that this is the smallest σ-algebra containing those elements.

This formalism of a countably-additive probability function on a countably-generated σ-algebra seems appropriate for epistemology. Even though finite beings are able to grasp infinite sets (we can think of the natural numbers, and the real numbers, and have propositions about each), it seems that we do so through finite means and this requires all the relevant infinite sets to be countably generated. We can grasp the integers through the concept of repeated addition and the rational numbers through division of integers. We can grasp the real numbers in a way, because each real number is the limit of a sequence of rationals and there are rational numbers that separate any pair of reals. But it is hard for us to grasp arbitrary sets of real numbers.

To specify an entire countably-additive probability on a countably-generated σ-algebra, it is sufficient to specify the probabilities of every finite intersection and finite union of the countably many generators, and this is the sort of thing that a finite mind can do through defining a computational process. For instance, the Borel sets are the σ-algebra of sets of real numbers generated by the open intervals. This algebra is generated by the countable collections of open intervals with rational endpoints. And standard Borel measure on the Borel sets is defined by stipulating that each open interval has measure equal to the difference of its endpoints. One doesn't need to directly specify the measures of each of the uncountably many Borel sets—specifying the measure of the countably many generators (and their unions and intersections) is sufficient, if the probability function is assumed to be countably additive. But for mere finite additivity, more would be needed, and this, I claim, would be out of the grasp of finite doxastic subjects. This is why other measures, like distance and mass, are typically assumed to satisfy countable additivity as well. (A version of this argument is given by Easwaran and Nielsen (2025).)

5.3.3 Arguments for countable additivity

The argument so far is purely mathematical. It shows that countably additive probability over countably generated σ-algebras has some nice features that make it seem epistemologically appropriate. But it doesn't show that the pragmatic or alethic role of a doxastic state requires countable additivity, in addition to non-negativity, normality, and (finite) additivity.

In my 2013b, I gave some more detailed arguments. The first argument (which comes from Skyrms (1992)) fits naturally with the pragmatic role for the doxastic state. In the Dutch book argument, we assumed that the agent was willing to buy any bet offered at a price below the probability and sell any bet offered at a price above the probability, but as mentioned in Section 5.2.1 we didn't make any assumptions about what they would do precisely *at* the probability. We can show that if an agent posts odds implying any collection of prices that satisfy countable additivity, there is no set of bets bought below the price or sold above the price that collectively lead to a Dutch book, while if they post odds implying a collection of prices that fail to satisfy countable additivity, then there is. Even though the collections of bets can be of any cardinality, the fact that they must be strictly favorable (and the fact that there must be a rational number between the favorable price and the posted price, and the fact that there are only countably many rational numbers) means that countable additivity is supported, not just finite additivity, and not full additivity. This argument has not been adapted into the context of the representation theorems, but I suspect a version of it could be.

The main argument however concerns a principle I called the "comparative principle". It said that there couldn't be two probability distributions over the same space, such that there is some partition in which every event has higher probability on the first distribution than the second. I showed that for every merely finitely additive distribution, there is some countably additive probability distribution that dominates it on some partition in this way. I suspect there may be a way to justify this principle from the alethic role for the doxastic state, though I haven't worked out the details. Seidenfeld et al. (2024) raise some challenges to this principle, giving one context in which no countably additive probability distributions exist, and another in which countably additive distributions are automatically fully additive, and they show that the version of the comparative principle I give for $\succ$ leads to full additivity if the ordering is linear. In Easwaran (2025), I show that the context in which there are no countably additive distributions is not one I am interested in (it is a context where many propositions that "should" be included are missing, so that certain collections of propositions appear to be partitions, even though they intuitively are not). I accept that in some contexts, countable additivity leads to full additivity (notably, it does so in contexts where there are only countably many possibilities!), and I accept that the ordering of propositions by comparative probability may not be a linear ordering. This deepens some of the worries for both the alethic and pragmatic foundations of degree of belief.

5.3.4 Infinitesimals?

Some authors have tried to respond to these worries by finding a mathematical structure other than the real numbers that might be able to represent degrees of belief and all the other concepts like accuracy, utility, and expected value. Abraham Robinson (1966) showed how to construct a field of "hyperreal" numbers that include the real numbers, but also infinitely small and infinitely large "numbers" as well. His particular construction relies on the Axiom of Choice, and it is meant to ensure that every statement expressible in first-order logic has the same truth value in the standard reals as well as in the hyperreals—for every two distinct hyperreals, there is a hyperrational between them; every positive hyperreal has a positive and a negative square root; and so on.

However, two of the important properties of the real numbers for some of their applications are *not* expressible in first-order logic. In particular, the field of hyperreals does not satisfy the Archimedean axiom or the axiom of continuity. No finite number of copies of an infinitely small hyperreal number can add up to more than a finite hyperreal number. There are bounded sets of hyperreals with no *least* upper bound. (For instance, the set of all infinitely small numbers has no least upper bound—if that least upper bound were itself infinitely small, then multiplying it by 2 would yield a larger infinitely small number, and if that least upper bound were *not* infinitely small, then dividing it by 2 would yield a smaller number that was still greater than all the infinitely small numbers.) Nelson (1977) has given an alternate framework to state versions of these axioms. Not *every* bounded set has a least upper bound—but every "internal" set does. There is no *finite* set of copies of some numbers that add up to more than another—but there is a "hyperfinite" set.

I have argued that the dependence on the Axiom of Choice makes it impossible for physical beings to have degrees of belief with the same structure as these hyperreals (Easwaran, 2014b). This field has *too many* elements for any physical fact about a subject to pick out one of them as the appropriate representation of a degree of belief. This is the opposite problem from the real numbers, which had *too few* elements to assign a different degree of belief to the empty set and to the proposition that the speed of light is a specific real number.

I think there is more interest in pursuing a slightly different strategy. Even without appealing to the Axiom of Choice, one can construct non-Archimedean fields that extend the real numbers. For instance, we can consider the set of all expressions consisting of one polynomial in ϵ divided by another, like

$$\frac{\epsilon}{1}, \frac{\epsilon^2 + 2\epsilon - 3}{\epsilon + 1}, \frac{\epsilon + 1}{\epsilon - 1}, \frac{\epsilon^3 + 2\epsilon^2 + 3\epsilon + 4}{\epsilon^2 - 1}.$$

Every real number occurs among these expressions (as a formal fraction where the numerator and denominator are multiples of the same polynomial of ϵ), but there are also more. It is not hard to define a linear ordering on these expressions subject to the constraint that $\frac{\epsilon}{1}$ is less than any positive real number, but greater than 0, and show precisely how this extends all the other elements of the field. This linear ordering behaves appropriately with respect to the arithmetical operations of addition and multiplication. Any such field will still violate the Archimedean and continuity axioms and will also have some first-order expressible properties that differ from the standard real numbers. In this particular field, there is no square root of ϵ.

I don't think that there is any particular reason to use this particular field to represent the degrees of belief of a particular subject. But David Hilbert (1899/1902) showed how to construct particular non-Archimedean fields to represent the coordinates of points in a non-Archimedean geometry. And I suspect that with a slight modification of the axioms for either a pragmatic or alethic foundation for degree of belief, a similar construction would be possible.

RESEARCH QUESTIONS

1. Is there a mathematical structure that enables finer representations than the real numbers, but not so fine as to get the problems of the hyperreals? (See McGee (1994) and Halpern (2009) for discussions of several options in the literature.)

2. Can an alternate axiomatization for either pragmatic or alethic representation theorems be given that gives rise to a particular non-Archimedean field in which the representation occurs?

3. How do the results change if we do or don't assume that for every pair of distinct degrees of belief, one is greater than the other?

4. What mathematical theory for conditional probability adequately deals with probability zero? (See Easwaran (2019) for more on this.)

CHAPTER 6

Rational change of belief

SO FAR, WE have focused on the aspects of Bayesianism dealing with what degrees of belief are, how they are measured, and what they should be like at a time. There has been some discussion of how they change over time, particularly at the end of Chapter 4, but this chapter goes into greater detail about this, in ways that help reveal the power and significance of Bayesian reasoning for quantitative work in the sciences.

Before we begin, I introduce some terminology. Norms involving a subject's doxastic state at a single time are said to be "synchronic". Norms regarding doxastic states at multiple times are said to be "diachronic". They concern how one's doxastic state should change or stay the same. However, there are related synchronic norms relating one's doxastic state at a time, and one's *plans* for future doxastic states, or one's *memories* of past doxastic states. These are in fact synchronic, because the plans, memories, and state that are being related are all at the same time. But because these plans and memories somehow point to other times, I will call these norms "quasi-diachronic".

In Section 3.5, we considered the idea that there might be a distinctive role for doxastic states grounded in appropriately responding to evidence. Most Bayesians have thought that the diachronic and quasi-diachronic norms described in this chapter are sufficient to do all the work that an evidential grounding is meant to do. In Section 6.2, I will demonstrate some of the power of these norms to accommodate many of our evidential intuitions. These diachronic and quasi-diachronic norms tell us how evidence supports theories, and do most of the more substantive work for Bayesianism in statistical reasoning and the philosophy of science. In this chapter, I will discuss these norms and some of their implications. In the following chapter, I will investigate the

DOI: 10.4324/9781315626918-6

differences between the source of these norms and some of the other norms of Bayesianism.

6.1 CONDITIONAL PROBABILITY

All the diachronic and quasi-diachronic norms that have been proposed involve in some way the mathematical concept of conditional probability. Formally, if the probability of a proposition a is notated "$P(a)$", then the conditional probability of a proposition a *given* another proposition b is notated "$P(a|b)$". Mathematically, conditional probability is defined by $P(a|b) = P(a \cap b)/P(b)$, whenever $P(b) > 0$.[1] One useful metaphor to keep in mind is that the conditional probability of a given b represents the "fraction" of b possibilities that are also a possibilities. However, it will be more helpful to understand this concept to see the various roles that it is philosophical said to play, which are discussed over the course of this chapter.

One strand of literature that I will not discuss much is the question of whether conditional probability is the probability of a conditional. Some have said that $P(a|b)$ is what we should think of as the probability of the conditional proposition, "If b then a" (Adams, 1965). However, it is notoriously difficult to say what the meaning or truth conditions of a conditional proposition are (Bennett, 2003). Furthermore, it has been shown that if the logic of conditionals satisfies certain plausible principles, then the equation of the probability of all conditionals with the corresponding conditional probability is impossible unless all probabilities are 0 or 1 (Lewis, 1976). Some, like Edgington (1995), have used this to argue that conditionals are not the sort of propositions that are true or false, and that they instead function in other ways. Some recent results showing just how much of the logic of conditionals can be saved with this sort of move are due to Bacon (2015). But I will not discuss this idea further. I will just consider propositions as sets of possibilities, and not worry about whether some of them might be "conditional" in some sense.

Hájek (2003) argues that we should give some account of what conditional probability is independent of its mathematical formulation in

[1] As I suggested in the last chapter, there are many relevant propositions for which $P(b) = 0$. There are a variety of philosophical and mathematical responses to this issue, which are mostly beyond the scope of this book. I discuss these approaches at greater length, and greater mathematical detail, in Easwaran (2019).

order to justify this formulation. I will do some of this in Section 6.3. But I think it will be helpful to go through some of its uses first.

6.2 CONDITIONALIZATION AND CONFIRMATION

The primary diachronic norm that has been proposed for Bayesianism is conditionalization. If P_0 is the probability function that represents a subject's degrees of belief at time t_0 and P_1 represents the subject's degrees of belief at a later time t_2, and if b is everything that the subject learns between t_0 and t_1, then the subject should have $P_1(a) = P_0(a|b)$. In Section 6.3, when I give the justifications for this norm, I will say more about what it means to say the subject "learns" b, and what the force of this norm should be. But before doing that, I will demonstrate some of the power of conditionalization in application.

In most of what follows, 'h' will stand for some hypothesis about the world, and 'e' will stand for some potential evidence that could be relevant to that hypothesis. Importantly, in this context, e is always itself a proposition. We refer to the degrees of belief before the update as the "prior probabilities", and the degrees of belief after the update as the "posterior probabilities". The idea is that the difference between the prior and posterior probability of h can tell us something about how much e "confirms" or "disconfirms" h. Much traditional Bayesian philosophy of science, from Hosiasson-Lindenbaum (1940) to Howson and Urbach (1989), focuses on these concepts.

6.2.1 Confirmation by consequences

First, we can see how a theory that makes a prediction can be confirmed by observing that prediction to be true, and that it is confirmed more strongly if the prediction is a more surprising one. That is, we consider the case where h entails e. In this case, there is no doxastic possibility for the agent on which h is true without e also being true (that is, h is a subset of e), so h and $h \cap e$ are equivalent for the agent. Thus, if the agent learns e (and nothing else) between t_1 and t_2, then these norms say that

$$P_2(h) = P_1(h|e) = \frac{P_1(h \cap e)}{P_1(e)} = \frac{P_1(h)}{P_1(e)}.$$

What this means is that if a hypothesis makes a prediction, and that prediction is learned to be true, then as long as one wasn't already quite confident in that prediction (as long as $P_1(e) < 1$), then one's degree of belief in the hypothesis should go up. In particular, one's degree of belief in the hypothesis should go up directly in proportion to how surprising the prediction was that was verified. An astronomical theory that successfully predicts that there will be a solar eclipse tomorrow will thereby receive greater confirmation than an astronomical theory that successfully predicts that the sun will rise tomorrow.

6.2.2 Confirmation by entailment

Conversely, consider what happens if e is some evidence that is strong enough to entail the hypothesis. There is no doxastic possibility where the evidence holds while the hypothesis does not, so $h \cap e$ is equivalent to e, so

$$P_2(h) = P_1(h|e) = \frac{P_1(h \cap e)}{P_1(e)} = \frac{P_1(e)}{P_1(e)} = 1.$$

Although "conditionalization" is often used to refer just to the mathematical norm about degrees of belief that we have already mentioned, it is natural to propose a parallel norm for the way that doxastic possibilities should change over time. When one learns a proposition e, one should eliminate all doxastic possibilities in which e doesn't hold. The remaining set of doxastic possibilities will not in general have probability equal to 1, so the norm of conditionalization then says to proportionally raise the probability of everything that is left in just such a way that the total probability is equal to 1. Eliminating the possibilities in which e doesn't hold is done by replacing h with $h \cap e$, and the proportional raising is done by dividing by $P(e)$. This is sometimes described by a metaphor of degree of belief as a substance that is spread out over the doxastic possibilities, with updates proceeding by removing the degree of belief from possibilities that are eliminated and spreading it out proportionally over what is left. (At least since Jeffrey (1965), many authors have been concerned about the fact that this update norm assumes that evidence is learned with certainty—alternatives that don't make this assumption will be discussed in Section 6.4.1.)

6.2.3 The justification of induction

Many further consequences of this norm of updating are explored by Howson and Urbach (1989). One particularly striking consequence is the way that this norm seems to vindicate a kind of inductive reasoning against Humean skepticism. Let h be some hypothesis, and let

$e_1, e_2, e_3, \ldots$ be a sequence of potential observations that would all be entailed by the truth of this hypothesis. (For instance, h might be the hypothesis that fire is always hot, and each e_i might be the claim that fire is hot on the ith time that we feel it.) Hume famously said that since there are no necessary connections between distinct matters of fact, i.e., we can always imagine possibilities in which the particular sequence of observations we have made is followed by some completely surprising and unrelated event, the observations $e_1, \ldots, e_i$ give us no reason to believe either h or even e_{i+1}.

However, the Bayesian says that we should think in terms of degree of belief rather than belief. It is clear that no number of observations of fire being hot can force us to be *certain* that fire will still be hot on our next observation. There is always room for a possibility on which the next observation is not hot. But as long as we had some slightly positive degree of belief in h to begin with, we can show that there must be some number i of observations of fire such that observing it to be hot i times leads us to have degree of belief .99 that the next observation will be hot as well. If this were not the case, then we could derive a violation of the Bayesian norms.

Let P_0 be the agent's degrees of belief before any observations and let P_i be the agent's degrees of belief after having observed $e_1, e_2, \ldots, e_i$. Since h entails every piece of evidence, we know that for any i, we must have $P_0(h) \leq P_0(e_1 \cap \cdots \cap e_i)$. But an application of the two norms of conditionalization tells us that,

$$P_1(e_2) = P_0(e_2 \cap e_1)/P_0(e_1),$$

so that

$$P_0(e_2 \cap e_1) = P_1(e_2)P_0(e_1).$$

Similarly,

$$P_0(e_3 \cap e_2 \cap e_1) = P_2(e_3)P_1(e_2)P_0(e_i),$$

and in general

$$P_0(e_1 \cap \cdots \cap e_i) = P_{i-1}(e_i)P_{i-2}(e_{i-1}) \cdots P_1(e_2)P_0(e_1).$$

That is, the prior degree of belief in a conjunction should be the product of the later degree of belief one has in each conjunct after learning all the prior conjuncts. If every single one of these later degrees of belief were below 99%, then for any positive number x, there is an i such that the product of the first i of these later degrees of belief is below x.

In particular, if $x = P_0(h)$, then this i would contradict the claim that $P_0(h) \leq P_0(e_1 \cap \cdots \cap e_i)$. The same argument will work for any threshold other than 99% as well. As long as there is some general hypothesis that entails a sequence of observations, and one starts out with positive degree of belief in that general hypothesis, then for any desired degree of belief, there is some number of these observations that together will entail that one should be that confident in at least the next observation.

This result has some limitations. First, the result doesn't tell us that $P_i(h)$ itself will ever be high. But this is reasonable. For instance, h might be the Newtonian theory of universal gravitation, which has implications about falling objects, the effect of the moon on the tides, and planetary motions. No matter how many observations of falling objects we observe on Earth, if we don't do any measurements of the correlation of the moon and the tides, and don't do any celestial mechanics, we would naturally maintain some uncertainty about these other consequences of h, and this will put some upper bound on how confident we can become in h. It doesn't even tell us that there is some N such that $P_n(e_{n+1})$ will always be high for any $n > N$. After all, if $e_1, \ldots, e_n$ are all observations of falling objects on Earth, while e_{n+1} is an observation of planetary motion, we might retain some uncertainty in it, no matter how many observations we have made on Earth. Instead, it just shows that for any such sequence of observations, and any degree of belief, eventually, *almost all* of the $P_n(e_{n+1})$ must be that high.

Second, the result doesn't apply if $P_0(h) = 0$. Karl Popper famously used a version of this argument in reverse (in Appendix *vii of his (1959)) to argue that $P_0(h)$ in fact *must* be 0 if h is a hypothesis that implies infinitely many logically distinct possible observations. He started from the Humean idea that no number of observations could ever give any reason to believe that the next observation will be similar. This means that in the above sequence, $P_n(e_{n+1})$ must never be high. But then by multiplying these together, and seeing that this is an upper bound to $P(h)$, we see that $P(h)$ must be 0. But Popper's argument depends on the assumption that if there is no "reason to believe" some proposition, then the probability of that proposition must not be high. This might be plausible for a kind of logical interpretation of probability, which is the only one that Popper himself considered.

But Bayesianism is a theory of degree of belief for particular doxastic subjects. In order to play either the pragmatic or the alethic role effectively for the subject, the doxastic state must include some degree of belief for each of the propositions that is relevant for the subject.

Inevitably, some of these will be high. There may be no *reason* why *particular* ones are high. But this just means that different subjects might have high credence in different propositions. The Bayesian rejects the inference from the fact that there is no reason for a particular degree of belief to be high to the conclusion that there *is* a reason for the degree of belief *not* to be high.

Third, there may well be many incompatible hypotheses that make incompatible predictions about particular potential observations. Goodman (1955) considered a case where h is the hypothesis that all emeralds are green, and each e_i is the observation that the ith emerald is green. But he also considered the "grue" hypothesis h', that the first N emeralds are green and all other emeralds are blue. Note that e_{N+1} is entailed by h, but $\bar{e}_{N+1}$ is entailed by h'. It can't be the case that P_N makes both e_{N+1} and $\bar{e}_{N+1}$ high. Any particular prior P_0 will yield a particular value for the posterior of e_{N+1}. Most of us likely have a prior that favors h over h', as long as our evidence is compatible with both. Goodman was interested in the question of whether there could be anything that justifies having such a prior. Nothing in Bayesian theory does justify having such a prior. But also, nothing rules it out.

6.2.4 The paradox of the ravens

Hempel (1945) raised a paradox for theories of confirmation. Part of his project was to axiomatize a theory of confirmation as a relationship between evidence and hypothesis. He did not assume a Bayesian account, but wanted this concept of confirmation to be a type of logical relation. Two features he thought this relation should have are that a hypothesis of the form "all a's are b" should be confirmed by any observation of an a that is b, and that if two hypotheses are logically equivalent, then any evidence that confirms one should confirm the other. But this leads to an apparent paradox.

The hypothesis "all ravens are black" is equivalent to the hypothesis "all non-black things are non-ravens". The first principle says that the former should be confirmed by the observation of a black raven. But it also says the latter should be confirmed by the observation of a white shoe. Hempel thought it was implausible that the observation of a white shoe could confirm the hypothesis that all ravens are black. So at least one of these principles had to be false.

Hosiasson-Lindenbaum (1940) gave one resolution of the paradox by working within a Bayesian framework of confirmation. She measured the confirmation of a hypothesis by evidence by looking at the change in probability from the prior to the posterior. This analysis of confirmation automatically satisfies the second principle, as long as the subject is certain that the hypotheses are equivalent. She assumed that the observation would be made by a random selection from the universe of objects, and that even when one is uncertain that the hypothesis is true, one starts with the prior that the sampled object is much more likely to be non-black than it is to be a raven. In this case, she was able to demonstrate that the observation of the black raven raises the probability of the hypothesis by more than the observation of the non-black non-raven (for reasons related to the fact already observed, that surprising evidence provides more confirmation than unsurprising evidence). She also provides another route to this result by noting that our prior that other ravens are like a given raven is higher than our prior that other non-black things are like a given non-black thing. She doesn't argue that either of these two features of the prior is *objectively required*, just that these are natural thoughts about our "present state of knowledge".

This response is notable because it makes use of the quantitative features of Bayesian confirmation. When Hempel had considered a relation of confirmation that is either present or absent between two propositions, it seemed that one would want the relation to be completely absent between the white shoe and the hypothesis that all ravens are black. But once the relation is numerical, we can dissolve the problem by noting that one is much less than the other. This is related to the fact that the Bayesian framework also allows us to draw distinctions of degree of belief that are finer than belief. If we accept the Lockean thesis, mentioned in Section 1.5.1.2, we can allow that there are propositions that a subject believes by having a degree of belief just above the threshold, and others that are believed by having a degree of belief that is almost 1.

As noted in Section 4.1.3.3, there are many different numerical representations of degree of belief, not just the one that satisfies the probability axioms. Meacham and Weisberg (2011) worried that this would make it hard to say what it means for one degree of belief to be "much greater" or "slightly greater" than another, since different representations might put different degrees of belief closer or farther on the numerical scale. Fitelson (1999) compares many of the different quantitative measures of confirmation that exist. But because Hosiasson-Lindenbaum's case involves different pieces of evidence acting on a single hypothesis, it turns out that her result is completely robust across all measures of confirmation.

While this seems like an adequate resolution of the paradox of the ravens from a Bayesian perspective, more has been done. Swinburne (1971) notes that the Bayesian account actually does reject the claim that the hypothesis that "all *as* are *b*" is always confirmed by an observation of an *a* that is *b*. When we look at examples, we can see how this might naturally fail. You probably have a fairly high degree of belief that all humans are less than 8 feet tall and a high degree of belief that all unicorns are pink. (The former because you have some sense of the distribution of heights of people and the latter because you have high degree of belief that there are *no* unicorns at all, so that all *zero* of them are pink.) However, if you were to observe a person whose height was $7'11''$ or observe an actual unicorn that was pink, your degree of belief in these hypotheses would likely go down. If there is a person who is that close to 8 feet, then there could easily be another who was just a little taller; if there is one unicorn of any color, then there could be others that wouldn't all necessarily be the same color.

Although we end up giving up on a nice simple logical formulation of the concept of confirmation, we are able to get a flexible formalism that enables us to talk generally about confirmation.

6.2.5 Bayes' Theorem

There is a mathematical result in probability theory known as "Bayes' Theorem" that is often seen as giving the closest thing to a simple logical formulation of the concept of confirmation as is possible in a Bayesian framework. In fact, it is precisely because this theorem has been seen as so central to the framework that the framework is named "Bayesianism". This theorem was first proved by Bayes (1763) in one of the first explicit discussions of probabilistic confirmation by observations.

The basic observation behind the theory is that $P_0(h \cap e)$ can be related to two different conditional probabilities. On the one hand, $P_0(h \cap e) = P_0(h|e)P_0(e)$. On the other, $P_0(h \cap e) = P_0(e|h)P_0(h)$. This is because of the symmetry of the way in which e and h enter into the formula. As long as the values involved are non-zero, we can put these two equations together to get Bayes' Theorem:[2]

$$P_0(h|e) = P_0(e|h)\frac{P_0(h)}{P_0(e)}.$$

[2]It is also true that $P_0(e|h) = P_0(h|e)\frac{P_0(e)}{P_0(h)}$, but this order turns out to be less useful.

Because the norm of conditionalization says that $P_1(h) = P_0(h|e)$, the left hand side of the equation represents the posterior of the hypothesis. The terms in the fraction on the right are the priors of the hypothesis and the evidence. And the term "$P_0(e|h)$" is called the "likelihood". Sometimes it is referred to as the likelihood of the hypothesis for the evidence or the likelihood of the evidence on the hypothesis. In ordinary language, we often use the term "likelihood" as a synonym for "probability", but it's important to note that in the context of Bayesian confirmation theory, it is a technical term that specifically refers to this conditional probability.

When comparing multiple hypotheses, $h_1, h_2, h_3, \ldots,$ we see that $P_1(h_i) = \frac{P_0(e|h_i)P_0(h_i)}{P_0(e)}$. The denominator of the fraction is the same for all hypotheses—it is just the prior of the evidence.[3] So the numerator does all the work. We can see that the posterior of each hypothesis is proportional to the prior of that hypothesis multiplied by the likelihood it gives the evidence. The highest likelihood a hypothesis can give to the evidence is 1, so we saw in Section 6.2.1 that a hypothesis that gives a likelihood of 1 is always confirmed. I think this is also what is behind the intuition that a universal generalization is confirmed by its instances—if we imagine that we already know we are about to observe a raven, or a unicorn, or a person, then the best we can do for the hypothesis that all ravens are black, or that all unicorns are pink, or that all people are under 8 feet is to observe that it is black, or pink, or under 8 feet. But in the case of the pink unicorn, it was the first step of learning that we were about to see a unicorn at all that was unlikely and disconfirmed the hypothesis. And in the case of the person, we learned more than just that they were under 8 feet—we also learned that they were very nearly 8 feet, which was less likely on this hypothesis than

[3]If one wants to investigate the denominator, note that if the h_i are a mutually exclusive and exhaustive set of possibilities, so that the agent is certain that exactly one of them is true, then $P(e) = P(e \cap h_1) + \cdots + P(e \cap h_n)$. This fact will be discussed further in Section 6.5.1. In each case, $P(e \cap h_i) = P(e|h_i)P(h_i)$, so this denominator can also be worked out given the likelihoods and priors of the various hypotheses. Many formulations of Bayes' theorem explicitly write this out, though I find that it is often not helpful.

However, in realistic scenarios, we are often considering a few explicit hypotheses, but are not certain that one of them must be true. Thus, we must add some additional "none of the above" hypothesis, which is traditionally called the "catch-all". Although likelihoods of explicit hypotheses can often be determined by studying the hypothesis itself, the likelihood for the catch-all is much less determinate. See for instance Maher (1995) for an approach to this issue.

the alternative. Bayes' Theorem lets us focus on precisely the aspect that is relevant, which turns out to be this general epistemic relation of likelihood, rather than a formal logical relation of being an instance.

As mentioned in Section 4.1.3.3, it is sometimes helpful to represent degrees of belief by their odds rather than their probabilities. That is, the odds of a hypothesis are given by $O(h) = \frac{P(h)}{P(\bar{h})}$. Because Bayes' Theorem applies just as well to $\bar{h}$ as to h, we see that

$$O_1(h) = \frac{P_1(h)}{P_1(\bar{h})} = \frac{(P_0(e|h)P_0(h))/P_0(e)}{(P_0(e|\bar{h})P_0(\bar{h}))/P_0(e)} = \frac{P_0(e|h)P_0(h)}{P_0(e|\bar{h})P_0(\bar{h})} = \frac{P_0(e|h)}{P_0(e|\bar{h})}O_0(h).$$

The quantity $\frac{P_0(e|h)}{P_0(e|\bar{h})}$ is often called the "likelihood ratio" (it is the ratio of the likelihoods of h and $\bar{h}$ for e), and so the odds formulation of Bayes' Theorem states that the posterior odds equal the prior odds multiplied by the likelihood ratio. As mentioned in Section 4.1.3.4, this means that we can think of the likelihood ratio as giving a measure of the confirmation that the evidence gives the hypothesis.

One reason this formulation is helpful is that in many cases, although different people have different priors for the various hypotheses, they often have the same likelihoods. This is because the hypothesis itself is often phrased in probabilistic terms, and people's likelihoods tend to match the probabilities stated in the theory. For instance, if our hypotheses are about the bias of a coin, then it is very natural to think that the bias of the coin for or against heads gives the subject's likelihood for that hypothesis.[4] Consider a situation where a subject is certain that a coin is either biased with a .6 probability of heads, or fair with .5 probability of heads. Then the likelihood ratio of a single observation of heads will be 5/6, and an observation of n heads will be $5^n/6^n$. If the subject has prior degree of belief 2/3 that the coin is fair, then their prior odds is 2. After observing one flip of heads, their posterior odds will be $10/6 = 5/3$, which results in a posterior probability of 5/8 that the coin is fair and 3/8 that it is biased with a 60% chance of heads.

For more complicated pieces of evidence, like those involved in statistical sampling, we can do the same sort of thing. Imagine that we flip the coin 100 times, and it comes up heads precisely 63 of the times. The mathematics of combinatorics defines a number "$\binom{100}{63}$" that represents

[4]Lewis (1980) argues that this is in fact a rational requirement on degrees of belief that goes beyond probabilism, but de Finetti (1974) argues that these "hypotheses" are not really about the physical world at all, but are just expressions of this particular feature of our degrees of belief.

how many different subsets of size 63 there are in a set of 100 elements. This tells us how many different sequences of 63 heads and 37 tails there are. We can calculate that the likelihood of this evidence on the hypothesis that the coin has a bias of .6 is $\binom{100}{63} \cdot .6^{63} \cdot .4^{37}$ and the likelihood of this evidence on the hypothesis that the coin has a bias of .5 is $\binom{100}{63} \cdot .5^{63} \cdot .5^{37}$. In this case, it turns out that the likelihood ratio is about 1/40, so if the prior odds of the coin being fair were 2, then the posterior odds are 1/20, representing a drop from a probability of about .666 to a probability of about .05. Observing 63 heads in a string of 100 flips is fairly good evidence that the coin is biased at the level of .6. But if someone had prior odds of 99, representing a prior credence of .99 that the coin was fair, then the posterior odds would be about 2.5—they would still be fairly confident the coin was fair. But observing another 63 heads in another string of 100 flips would provide new evidence whose likelihood ratio is about 1/40, and this would be fairly convincing to the subject.

This is exactly the logic behind political polling, and it shows why a sample of only a few hundred voters can often give us reasonably strong evidence about the division of opinion in the population at large. If (and this is a big "if"!) we can assume that every person in the population is equally likely to respond to the poll, then the hypothesis that 60% of the electorate prefers candidate A is going to make each response to the poll behave like a coin with bias .6, while the hypothesis that 50% of the electorate prefers candidate A is going to make each response to the poll behave like a coin with bias .5. As long as someone has priors about different distributions of support that aren't too strongly inclined in one direction or another, even a poll of just a few hundred voters would likely convince them about the distribution of opinion. The only difference the size of the population as a whole makes is to the assumption that the pollster is equally likely to sample everyone. (Though in recent decades, the bigger issue has been not about the size of the population, but rather the fact that the kinds of people who tend to answer the phone when pollsters call have very different political opinions from the kinds of people who tend not to answer the phone.)

Although Bayes' Theorem is very useful in discussing the updates of Bayesian subjects, it is just a theorem of synchronic probability, and just shows how to reverse the order of conditional probability by multiplying and dividing by the probabilities of the relevant propositions. Thus, it is used in any context where the mathematics of probability is used, including in contexts where no degrees of belief are involved, as in much of classical statistics. But within the Bayesian paradigm, where probabilities are interpreted as degrees of belief, it has many distinctive uses. By comparing the likelihoods of two different hypotheses for the same piece of evidence, we can see which it confirms more strongly. And by comparing the likelihood of one hypothesis for two different pieces of evidence, we can see synchronically which would confirm it more strongly, if, hypothetically, either were to be learned.

6.2.6 The problem of the priors

In a case with two hypotheses, the likelihood ratio for one hypothesis was just equal to the ratio of the likelihoods for the two hypotheses. Thus, if there is agreement about the likelihoods of the hypotheses, then there is agreement about whether the evidence confirms or disconfirms each hypothesis. Subjects with different priors might disagree about how much evidence it would take for them to become confident in a hypothesis on the basis of their priors (one of them might in fact even start out confident in the hypothesis even before gathering *any* evidence!) but they will agree about how powerful the evidence is. However, when there are three or more hypotheses, it turns out that subjects with different priors can disagree about whether evidence confirms or disconfirms a particular hypothesis.

Consider three hypotheses about the history of the earth. h_1 says that life on Earth has existed for hundreds of millions of years and that during that time there were many changes in the species that exist. h_2 says that life on Earth has existed for hundreds of millions of years but has been roughly similar throughout that entire period. h_3 says that life on Earth has existed for 6000 years, and it was created by a divine being, who also created layers of rock that simulate a long past in order to test the faith of humans.

Let e be the observation that some layers of rock contain the apparent skeletons of large reptiles that don't exist today. Further, let us assume that all researchers agree that $P(e|h_1) = .05$, $P(e|h_2) = .005$, and $P(e|h_3) = .5$. (The idea is that under the evolutionary theory, it is likely that there were different creatures in the past, but moderately unlikely that they were large reptiles; under the stable theory it is quite unlikely that any fossils of this sort would exist; and under the young earth creationist theory, it is just as likely as not that the deity would include this sort of fossil.)

For any researcher, Bayes' Theorem says that their posterior degrees of belief in each hypothesis will be proportional to the priors times the likelihoods. For someone with priors of $P(h_1) = .09$, $P(h_2) = .9$, and $P(h_3) = .009$ (with .001 left over for a catch-all), the posteriors will be proportional to $.09 \cdot .05$, $.9 \cdot .005$, and $.009 \cdot .5$. Because of how I have chosen these numbers, all three of these values are equal, so the posteriors of all three hypotheses will be equal. As long as the catch-all doesn't have a high likelihood for this evidence, each of these posteriors will be close to 1/3, so in particular, a person with this prior will see this evidence as confirmation for h_1.

However, for someone with prior degrees of belief on which $P(h_1) = .09$, $P(h_2) = .009$, and $P(h_3) = .9$ (with .001 left over for a catch-all), the situation will be quite different. Their posteriors will be proportional to $.09 \cdot .05$, $.009 \cdot .005$, and $.9 \cdot .5$. That is, while their prior for h_3 was 10 times their prior for h_1, their posterior for h_3 will be 100 times their posterior for h_1! Regardless of their likelihood for the catch-all, this means their posterior in h_1 will be less than .01. Thus, they see this evidence as *dis*confirmation for h_1.

Even though the two researchers agreed on the likelihoods, and had the same prior for h_1, one of them took the evidence e to be moderately strong in favor of h_1 while the other took it to be moderately strong evidence against h_1.

This is because the two gave different credence to various alternatives. The ratio of the degree of belief of any two hypotheses will change in the direction of the ratio of the likelihoods of those hypotheses. But even though both of them agreed that the evidence had a likelihood for h_1 that was between that of h_2 and h_3, the effect on their degree of belief in h_1 depended on how the rest of their prior was distributed between h_2 and h_3. The hypothesis with the highest likelihood will always go up in probability, and the hypothesis with the lowest likelihood will always go down in probability. But for hypotheses with intermediate likelihoods, it depends on how much of the prior was assigned to hypotheses with higher likelihood and how much to hypotheses with lower likelihoods.

Some statisticians, like Royall (1997), argue that we should think of the likelihoods themselves as measuring the full import of the evidence, rather than thinking in terms of confirmation or disconfirmation. But for Bayesians, confirmation and disconfirmation are important, and this depends on the priors. This is often said to be the "problem of the priors". Even if everyone agrees on the likelihood of the evidence on each hypothesis, whether the evidence counts for or against a particular hypothesis depends on the priors of the other hypotheses. Many philosophers have thought that some evidence should objectively count in favor of certain hypotheses and against others, so that this is a demerit for Bayesianism. However, I accept this conclusion. I will return to this point in Section 8.1.5, where I argue that even though there is no objectively correct way to respond to evidence, if we have some idea of what the person's prior might have been like, we can be pretty confident that they are disobeying their own concept of confirmation, if their update is very different from our own.

6.3 ARGUMENTS FOR CONDITIONALIZATION

Now that I have shown some of the consequences of conditionalization, I will return to the question of how to justify it. I will consider two pragmatic arguments and two alethic arguments.

6.3.1 The Dutch book for conditionalization

The basic Dutch book argument for conditionalization was first published by Teller (1973) (though he attributes the argument to David Lewis). We must first specify what role conditional degree of belief should play in governing betting behavior. As a reminder, in the Dutch book argument, the agent's degree of belief in a proposition is numerically represented as the price such that they are willing to buy a bet that returns \$1 if the proposition is true at any amount less than that price, and to sell such a bet for any amount greater than that price. Bets for other stakes have the fair price multiplied by those stakes.

Conditional degree of belief is represented by the price one finds fair for a *conditional* bet. A conditional bet on a, given b, is a contract that pays 1 if a and b are both true, pays nothing if a is false and b is true, but returns the original payment if b is false. The idea is that the bet is called off if the condition doesn't happen. Table 6.1 illustrates the net results of buying two different unconditional bets, and a conditional bet. Table 6.2 then shows what happens to the first two bets if we set the stakes of the bet on b precisely at $\frac{P(a \cap b)}{P(b)}$. (I will not here discuss the situation where $P(b) = 0$, because it gets quite mathematically complex. But it is discussed by Easwaran (2013b), Rescorla (2018), Nielsen (2021b), and Easwaran and Nielsen (2025).)

When we compare the results of the conditional bet on a, given b, to the sum of these two unconditional bets at these specific stakes, we see that these are precisely equal if $P(a|b) = \frac{P(a \cap b)}{P(b)}$, but that if these values are not equal, then these bets can be put together (along with a fourth

TABLE 6.1 Net outcomes for buying bets on b, on $a \cap b$, and on a given b, for stakes S_b, $S_{a \cap b}$, and $S_{a|b}$, precisely at the fair price

	$b \cap a$	$b \cap \bar{a}$	$\bar{b}$
Bet on b	$+(1 - P(b))S_B$	$+(1 - P(b))S_B$	$-P(b)S_B$
Bet on $a \cap b$	$+(1 - P(a \cap b))S_{a \cap b}$	$-P(a \cap b)S_{a \cap b}$	$-P(a \cap b)S_{a \cap b}$
Bet on a given b	$+(1 - P(a\|b))S_{a\|b}$	$-P(a\|b)S_{a\|b}$	0

bet on b) to yield a Dutch book. If $P(a|b) = \frac{P(a \cap b)}{P(b)} - \epsilon$, then we have the agent sell the conditional bet, buy the bet on $a \cap b$ and sell the bet on b, all within $\epsilon/4$ of the fair price. The net result will be a loss of at least $\epsilon/4$ in the b worlds, and neutrality in the $\bar{b}$ worlds. We can turn this into a guaranteed loss in all worlds if we further sell the agent a bet on b for stakes less than $\epsilon/4$, since this bet loses in the $\bar{b}$ worlds and wins less than $\epsilon/4$ in the b worlds. We can do the opposite if the difference goes the other direction.

This yields a justification of the synchronic norm that $P(a|b) = \frac{P(a \cap b)}{P(b)}$. But the more interesting point is to justify the diachronic norm of conditionalization. Since this involves the agent's degrees of belief at different times, it requires bets at different times, made according to the agent's degrees of belief at those times.

Consider an agent who learns only b between t_0 and t_1, and yet they have $P_1(a) > P_0(a|b)$. We could buy from them a conditional bet on a given b at t_0 for slightly more than $P_0(a|b)$, and then at time t_1 sell them a bet on a for slightly less than $P_1(a)$. If a is true, the winnings of both bets cancel and if it is false, there are no winnings. Either way, they have lost some money. If the person has $P_1(a) < P_0(a|b)$, then the opposite strategy yields a guaranteed loss. The only way to be immune to this strategy is if the person has $P_1(a) = P(a|b)$.

TABLE 6.2 Net outcomes for selling a bet on b at stakes $\frac{P(a \cap b)}{P(b)}$, buying a bet on $a \cap b$ at stakes 1, and doing both, precisely at the fair price

	$b \cap a$	$b \cap \bar{a}$	$\bar{b}$
Sell bet on b at stakes $\frac{P(a \cap b)}{P(b)}$	$-\frac{P(a \cap b)}{P(b)} + P(a \cap b)$	$-\frac{P(a \cap b)}{P(b)} + P(a \cap b)$	$+P(a \cap b)$
Buy bet on $a \cap b$ at stakes 1	$+(1 - P(a \cap b))$	$-P(a \cap b)$	$-P(a \cap b)$
Sum of both	$+(1 - \frac{P(a \cap b)}{P(b)})$	$-\frac{P(a \cap b)}{P(b)}$	0

However, we have to be careful about how we interpret this result. Christensen (1991) points out that a superficially similar-looking strategy seems to yield an argument that no agent should *ever* change their degrees of belief, *even* by conditionalization. Imagine that the agent has $P_0(a) < P_1(a)$. Then we buy from them a bet on a at t_0, and sell it back to them at t_1, for guaranteed profit.

The issue is the question of what is meant by a "guaranteed" profit. The latter profit is guaranteed if the agent does in fact increase their credence in a. But in most cases, the increase in credence is driven by something that the agent learned. And whatever it is that they learned, call it b, wasn't guaranteed to be true. (At least, they didn't see it as guaranteed—if they did, then they wouldn't have needed to learn it.) If they had learned the opposite, $\bar{b}$, and if their update in both situations obeys conditionalization, then they would have decreased their credence in a.[5]

So we should be slightly more careful with the initial Dutch book for conditionalization. Rather than just calling it "P_1", let's call it "Q_b", and assume that this is the doxastic state the agent comes to have in every world where b is true, and that there is some other doxastic state $Q_{\bar{b}}$ that they come to have in every world where b is false. (In this and the next several arguments, I will reserve "P" for probabilities related by conditionalization to P_0, and use "Q" for updates that may or may not be normatively correct.) Now, if $Q_b(a) = P_0(a|b)+\epsilon$, we can put together a Dutch book more carefully. We pay the person right now $P_0(a|b)+\epsilon/4$ for a conditional bet of stakes 1 on a, given b. We also sell the person a secondary bet on b at stakes $\epsilon/4$. And then after the person has learned whether or not b is true, if they have learned that b is true, we sell them a bet for $Q_b(a) - \epsilon/4$ of stakes 1 on a. In the b worlds, the winnings of the two bets on a cancel out, and the two main bets together guarantee a loss of $\epsilon/2$, while the secondary bet gives them winnings of less than $\epsilon/4$. In the $\bar{b}$ worlds, the conditional bet on a is called off, and the second bet on a never happens, but the person has bought a bet on b, so again they lose money.

This helps us better characterize precisely what situations are relevant for conditionalization. The strategy only works if there is some update that the person will do in all the b worlds, but not any of the

[5] That's because $P(a) = P(a \cap b) + P(a \cap \bar{b}) = P(a|b)P(b) + P(a|\bar{b})P(\bar{b})$. Since $P(b) + P(\bar{b}) = 1$, if both $P(a|b)$ and $P(a|\bar{b})$ were greater than $P(a)$, then the right hand side of that equation would be greater than $P(a)$, which is a contradiction.

$\bar{b}$ worlds. And the strategy itself involves some acts before the learning event that occur in all worlds, and some acts after the learning event that occur in all and only the b worlds. And this is reasonable. If an update happens in some set of worlds that is not all and only the b worlds, there's reason to think that the update is not driven by b. (We will consider some of these in Section 6.4.2.) And if the strategy involves anything before the learning event that is different in different worlds, or anything after the learning event that isn't the same in all b worlds, then the agent could use the fact of the strategy's implementation to learn something more about the world.

6.3.2 Good's theorem

One other result that provides a pragmatic justification for conditionalization is given by Good (1967). This argument has more often been taken as an argument that *presupposes* conditionalization and then argues that agents should gather information, but I will argue that it can be turned into an argument for conditionalization.

Good was considering the "requirement of total evidence", which is the requirement that one should in fact update on *all* the information one has, rather than part of it. He considers an agent who is deciding whether to do act A_1 or A_2, who has probability function P_0. But the person can, at no cost, gain the information of whether or not e is true, and update to P_e or $P_{\bar{e}}$ and then decide which act to do. Let's use $Exp_0(A_1)$ to represent the expected utility, computed according to probability function P_0 of A_1, and similarly for $Exp_e(A_1)$ and $Exp_{\bar{e}}(A_1)$ for P_e and $P_{\bar{e}}$, and similarly for A_2.

The overall structure of the argument is that he notes that if the agent acts before learning, then they have two possible strategies—do A_1 in every world, or do A_2 in every world. But if the agent uses the learning event to choose what to do, they have four possible strategies—the two above, as well as a strategy of doing A_1 in the e worlds and A_2 in the $\bar{e}$ worlds, and vice versa. He shows that if the agent's current doxastic state evaluates one of the latter two strategies as having higher expected utility than either of the ones available now, then updating by conditionalization and then choosing the act with highest posterior expected utility will in fact result in that optimal strategy. Thus, the agent should prefer now to wait before choosing.

The rest of this section gives the mathematical details of this argument, showing first that conditionalization leads to the choice of that optimal strategy, and second that for any update *other* than conditionalization, there is a pair of options A_1 and A_2 where that update leads to a strategy that the agent already evaluates as worse than one of the options available now. Readers uninterested in the mathematical details can skip these paragraphs.

First, he notes that $P_0(p) = P_0(e) \cdot P_e(p) + P_0(\bar{e}) \cdot P_{\bar{e}}(p)$ for any proposition (assuming the agent updates by conditionalization)—that is, P_0 is a "mixture" of P_e and $P_{\bar{e}}$. (This fact will be discussed further in Section 6.5.1.) But since the expected utility of an act according to a probability function is just the result of multiplying the utility of that act in any world times the probability of that world, this means that $Exp_0(A_1) = P_0(e) \cdot Exp_e(A_1) + P_0(\bar{e}) \cdot Exp_{\bar{e}}(A_1)$, and similarly for A_2.

Then, he notes that if the agent plans to learn whether or not e is true, and then choose between A_1 and A_2, they'll plan to choose whichever of the two has higher expected utility after updating. That is, in the e worlds, they'll choose A_1 if $Exp_e(A_1) > Exp_e(A_2)$ and A_2 if $Exp_e(A_2) > Exp_e(A_1)$, and similarly in the $\bar{e}$ worlds.

If they end up choosing A_1 in both sets of worlds, then the expected utility of this strategy is $P_0(e) \cdot Exp_e(A_1) + P_0(\bar{e}) \cdot Exp_{\bar{e}}(A_1)$, which is just $Exp_0(A_1)$. Similarly, if they end up choosing A_2 in both sets of worlds.

But if they choose A_1 in the e worlds and A_2 in the $\bar{e}$ worlds, then the expected utility of the strategy is $P_0(e) \cdot Exp_e(A_1) + P_0(\bar{e}) \cdot Exp_{\bar{e}}(A_2)$. Since they chose A_1 in the e worlds, this must mean that $Exp_e(A_1) > Exp_e(A_2)$, so $P_0(e) \cdot Exp_e(A_1) + P_0(\bar{e}) \cdot Exp_{\bar{e}}(A_2) > P_0(e) \cdot Exp_e(A_2) + P_0(\bar{e}) \cdot Exp_{\bar{e}}(A_2) = Exp_0(A_2)$. Similarly, since they chose A_2 in the $\bar{e}$ worlds, this must mean that $Exp_{\bar{e}}(A_2) > Exp_{\bar{e}}(A_1)$, so $P_0(e) \cdot Exp_e(A_1) + P_0(\bar{e}) \cdot Exp_{\bar{e}}(A_2) > P_0(e) \cdot Exp_e(A_1) + P_0(\bar{e}) \cdot Exp_{\bar{e}}(A_1) = Exp_0(A_1)$.

To summarize, if the person waits for the information, either it doesn't matter (in which case they do the same act regardless, and they get the same outcome they could have by not waiting for information) or it does matter (in which case they do a different act in each type of world, and end up with a better expected utility than they could have by locking in either of the two acts right away). Thus, waiting for the information is always at least as good, and sometimes better, if we assume the agent updates by conditionalization.

But this can be tweaked to be turned into an argument against any update other than conditionalization.

Assume that the person currently has doxastic state given by probability function P_0, and that P_e and $P_{\bar{e}}$ are the results of updating this function by conditionalization on e and $\bar{e}$ respectively. But assume that the person plans to update to Q_e in the e worlds and to $Q_{\bar{e}}$ in the $\bar{e}$ worlds, where these are different functions from P_e and $P_{\bar{e}}$.

Let A_1 and A_2 be some acts such that A_1 is preferred to A_2 on both P_e and $P_{\bar{e}}$, but A_2 is preferred to A_1 on both Q_e and $Q_{\bar{e}}$.[6] Because A_1 is preferred to A_2 on both P_e and $P_{\bar{e}}$, and because P_0 is a mixture of P_e and $P_{\bar{e}}$, the agent currently prefers A_1. But they can see that updating to the Qs will lead them to choose A_2 instead, regardless of what happens.

Note that this is not an argument that the person *should* conditionalize—it just shows that the *plan* they currently evaluate as best is to update via conditionalization. Every plan to update in a non-conditionalization way is a plan that leads them to some strategies that they currently disprefer, but planning to update via conditionalization leads them only to strategies that they find at least as good as the acts they can currently do, and sometimes allows them to carry out better strategies. This is thus an expected utility argument for a quasi-diachronic norm, rather than the dominance argument above for the truly diachronic norm.

6.3.3 Limitations of the pragmatic arguments

These arguments work as stated for a pragmatic agent who follows the Dutch book argument, and can be tweaked to fit Ramsey and Savage's framework, where acts and states are independent, and preferences are given by standard expected value. However, for the versions of decision theory that allow for act-state dependence, or for risk-aversion, there are a few worries.

Buchak (2012) has shown that on her risk-sensitive decision theory, Good's theorem actually fails. A simple counterexample is actually given by the Allais paradox that she used to motivate her risk-sensitive decision theory, as described in Section 2.4. Consider someone who prefers A to B and D to C in this slightly modified Allais case in Table 6.3. Let e be the proposition that the ticket drawn is in the range 1–10 (so that $\bar{e}$ is the proposition that the ticket drawn is in the range 11–100.) Now imagine

[6] Let a_{1e} be some subset of e that has higher probability on P_e than Q_e, let a_{2e} be some subset of e that has higher probability on Q_e than P_e, let $a_{1\bar{e}}$ be some subset of $\bar{e}$ that has higher probability on $P_{\bar{e}}$ than $Q_{\bar{e}}$, and let $a_{2\bar{e}}$ be some subset of $\bar{e}$ that has higher probability on $Q_{\bar{e}}$ than $P_{\bar{e}}$. Let A_1 be an act that yields utility $1/P_e(a_{1e})$ on a_{1e}, $1/P_{\bar{e}}(a_{1\bar{e}})$ on $a_{1\bar{e}}$ and 0 elsewhere, and let A_2 be an act that yields $1/Q_e(a_{2e})$ on a_{2e} and $1/Q_{\bar{e}}(a_{2\bar{e}})$ on $a_{2\bar{e}}$ and 0 elsewhere. Then P_e and $P_{\bar{e}}$ both give expected utility 1 to A_1 and a lower expected utility to A_2, while Q_e and $Q_{\bar{e}}$ both give expected utility 1 to A_2 and a lower expected utility to A_1.

TABLE 6.3 Slightly modified Allais choices

Act	Tickets 1–9	Ticket 10	Tickets 11–100
A	\$5 million	\$0	\$0
B	\$1 million	\$1 million	\$1
C	\$5 million	\$0	\$1,000,001
D	\$1 million	\$1 million	\$1,000,000

that they have one of the choices right now, and are given the option to either pick now, or to first learn whether e is true. They know that if they learn e, and update by conditionalization, they will now be facing a decision between a .9 probability of \$5 million with a .1 probability of nothing (in either A or C), and a guarantee of \$1 million (in either B or D). Depending on what preference they will have after learning e, either the choice between A and B, or the choice between C and D, leads to a problem.

If they would prefer the guaranteed million with that information, then if they are currently facing a choice between A and B, they know that before learning the information, they would prefer A, but afterwards they would prefer B regardless of which they learn. If they prefer the .9 probability of \$5 million, then if they are currently facing a choice between C and D, they know that before learning the information, they would prefer D, but afterwards they would prefer C regardless of which they learn.

But Buchak argues that this is a good result. She argues that this is actually the phenomenon of faith, where it can be rational to take an act right now, rather than ask for more information before deciding whether to do so. The faithful subject suspects that gathering this information would distract them from the global risk properties of the situation they face right now. Buchak still accepts the rule of update by conditionalization. But she bases it on the diachronic Dutch book, even though it runs into this tension with Good's theorem. A faithful agent would prefer not to learn.

Since the proof of Good's theorem requires the factorization of expected utility into expected utility conditional on e and on $\bar{e}$, but causal, evidential, and functional decision theories each choose acts according to a slightly modified formula, it seems that there may be some problems for Good's theorem in these frameworks as well. But I am not aware of any specific challenges that have been raised on this basis.

However, Fusco (2025) notes that, for the decision theories on which the agent's own act is a proposition that they could update on, there is a core problem with the diachronic Dutch book argument. In particular, if the agent believes that the act is under their control, then their bets on this act won't have the same relationship to the price at all stakes. If I see a \$20 bill on the ground within reach, I expect that I'm likely to pick it up (provided there's nothing too dangerous or disgusting around it). So if you offer me a bet at \$1 stakes that I won't pick up the \$20 bill, I'm not going to take it, even at relatively favorable odds. However, if you offer me a bet at \$100 stakes that I won't pick up the \$20, suddenly I'll be willing to take that bet even at not-so-favorable odds—and if I take the bet, then I won't pick up the \$20, even though it's free money (because it will cost me in the bet). Fusco shows that a modified version of the Dutch book argument actually leads to a different sort of update, called "imaging", when the proposition learned is one of the agent's own acts.

6.3.4 The expected accuracy argument for conditionalization

The most prominent argument for updating via conditionalization in the alethic framework is that given by Greaves and Wallace (2006). Like in the rest of the alethic framework, they assume that accuracy is measured by a proper scoring rule $I(P, w)$ giving the inaccuracy of P in world w—propriety means that $Exp_P I(Q, w) > Exp_P I(P, w)$ for any distinct probability functions P and Q. They consider a doxastic subject who is about to learn whether or not e is true, and who wants the update plan that will minimize expected inaccuracy. They don't justify the use of expectation in the alethic setting, but I think this can be done, as argued in Section 3.6.1.

Let's say this subject currently has doxastic state given by probability function P_0. Let P_e and $P_{\bar{e}}$ be the probability functions that result from conditionalizing P_0 on e and $\bar{e}$, respectively. And let's say that the subject plans to update to Q_e in the e worlds and $Q_{\bar{e}}$ in the $\bar{e}$ worlds. Assume that there are finitely many worlds. (The argument in the case where there are infinitely many worlds is given by Easwaran (2013a), Nielsen (2022), and Easwaran and Nielsen (2025).) Then the expected inaccuracy of the subject's plan is given by

$$\sum_{w \in e} P_0(w) I(Q_e, w) + \sum_{w \in \bar{e}} P_0(w) I(Q_{\bar{e}}, w).$$

When $w \in e$, we have $P_0(w) = P_0(e)P_e(w)$, and when $w \in \bar{e}$, we have $P_0(w) = P_0(\bar{e})P_{\bar{e}}(w)$. Thus, this expected inaccuracy can be rewritten as

$$P_0(e) \sum_{w \in e} P_e(w) I(Q_e, w) + P_0(\bar{e}) \sum_{w \in \bar{e}} P_{\bar{e}}(w) I(Q_{\bar{e}}, w).$$

But $\sum_{w \in e} P_e(w) I(Q_e, w) = Exp_{P_e} I(Q_e, w)$ and $\sum_{w \in \bar{e}} P_{\bar{e}}(w) I(Q_{\bar{e}}, w) = Exp_{P_{\bar{e}}} I(Q_{\bar{e}}, w)$. Because I is a proper scoring rule, these expected inaccuracies are minimized when $Q_e = P_e$ and $Q_{\bar{e}} = P_{\bar{e}}$.

Thus, out of the plans that involve one update in the e worlds and another in the $\bar{e}$ worlds, the plan that minimizes expected inaccuracy is conditionalization. Like the argument based on Good's theorem, this argument uses expectations to give a justification of a quasi-diachronic norm of planning to conditionalize.

6.3.5 The accuracy dominance argument for conditionalization

Briggs and Pettigrew (2016) give a truly diachronic alethic argument for conditionalization (though Nielsen (2021a) identifies a few errors in the case where a proposition has probability 0 and fixes them). They consider a subject who at time t_0 has doxastic state Q_0, and then goes through a learning process where they will come to one doxastic state, Q_e in all the e worlds, and another, $Q_{\bar{e}}$ in all the $\bar{e}$ worlds. (When talking about the doxastic state after the update, without knowing whether it is in an e world or a $\bar{e}$ world, I will refer to it as "Q_1".) Because this is a truly diachronic argument, the subject is interested in minimizing their inaccuracy across times. In particular, Briggs and Pettigrew assume that the subject wants to minimize $\alpha I(Q_0, w) + \beta I(Q_1, w)$, for some positive constants α and β.

What they prove is that if Q_e and $Q_{\bar{e}}$ are not the probability functions that result from conditionalizing Q_0 on e and $\bar{e}$ respectively, then there exists a triple of probability functions, $P_0, P_e, P_{\bar{e}}$, where P_e and $P_{\bar{e}}$ *are* the result of conditionalizing P_0 on e and $\bar{e}$ respectively, such that $\alpha I(P_0, w) + \beta I(P_1, w) < \alpha I(Q_0, w) + \beta I(Q_1, w)$ at every world w.

First, they note that if one or more of $Q_0, Q_e, Q_{\bar{e}}$ are not probability functions, then the accuracy dominance arguments of Chapter 3 show that they are accuracy dominated. And since Q_e is only had in e worlds, and $Q_{\bar{e}}$ is only had in $\bar{e}$ worlds, it is not hard to show that the relevant Q is accuracy dominated if $Q_e(e) < 1$ or if $Q_{\bar{e}}(\bar{e}) < 1$. Thus, we can assume that all three are probability functions, and that the two possibilities for Q_1 give probability 1 to the respective piece of evidence.

At this point, they note that if $Q_0, Q_e, Q_{\bar{e}}$ are probability functions and $Q_e(e) = Q_{\bar{e}}(\bar{e}) = 1$, then Q_e and $Q_{\bar{e}}$ are the results of conditionalizing Q_0 on the relevant propositions if, and only if, Q_0 is a mixture of Q_e and $Q_{\bar{e}}$. Recall that for one probability function to be a mixture of others is for there to exist constants $k_e, k_{\bar{e}}$, such that $Q_0(p) = k_e \cdot Q_e(p) + k_{\bar{e}} \cdot Q_{\bar{e}}(p)$. (We can tell that this must be true by noting that $Q_e(e) = 1$ and $Q_{\bar{e}}(e) = 0$, so plugging e in for p yields that $k_e = Q_0(e)$, and similarly $k_{\bar{e}} = Q_0(\bar{e})$, and applying the definition of conditional probability.)

Because the property of one function being a mixture of two others is so much like the probability of a disjunctive proposition being the sum of the probabilities of two others, it turns out that the aspect of the Predd et al. (2009) result that proves that violations of additivity must lead to dominance translates perfectly to this new setting and shows that violations of the mixture principle must lead to dominance. Thus, if the Qs are not probability functions related by conditionalization, there is a triple of Ps that give strictly less total inaccuracy in every world.

Like the Dutch book argument, this is an argument for a truly diachronic norm of conditionalization, and it is based on dominance, rather than expected value.

6.3.6 Challenges to the diachronic arguments

From both the pragmatic and alethic roles for the doxastic state, there is an expected value based argument for a quasi-diachronic norm of *planning* to conditionalize, and a dominance argument for a truly diachronic norm of conditionalization. However, there are worries about whether any of these arguments do what we need an argument for conditionalization to do.

A quasi-diachronic norm just shows that conditionalization is the best plan. It doesn't show that, when it comes time to conditionalize, the subject should actually go through with the plan. Paul (2014b) argues that there may not be a general reason to follow through.

The truly diachronic norm would give something like this. But it raises other worries. For one, Christensen (1991) argues that these diachronic arguments only work if the stages of the self at different times really do have a rational requirement of coherence. Merely showing that there is a guaranteed loss is not enough unless the loss is shared by something with a real rational unity to it. He argues that the diachronic Dutch book against violations of conditionalization is a lot like a similar

Dutch book he raises against having *any* difference in degree of belief with one's spouse, if the two of you share a bank account. After all, if two people who share a bank account have different degrees of belief, then one will be willing to buy a bet at a higher price, and the other to sell the same bet at a lower price, and regardless of how the winnings turn out, the bank account has a net loss. He notes that this is a bit of a tragedy, but it is not a failure of rationality—he claims there is no rational requirement to have the same degrees of belief as one's spouse on everything, even when there is a shared a bank account.

But I say we should be more careful—there is a rational requirement not to share your finances with someone if you're not willing to plan and work together to protect those finances. This is one of the kinds of considerations that motivated me to include groups or "associations" in my list of doxastic subjects that may be subject to these norms.

I might prefer the bedroom painted blue while my partner prefers the bedroom painted red. That means that in *some* sense, I would be willing to pay someone to paint the bedroom blue, and my partner would be willing pay someone to paint the bedroom red. However, this preference doesn't translate into direct action the way we might naively assume. If we both acted in these ways, then of course we could have money extracted from us by a clever painter. Are we just lucky that no enterprising painter is quick and efficient enough to paint the bedroom cheaply enough to get each of us to pay in sequence?

No! Part of sharing finances is sharing a commitment to one another not to make purchases that might specifically displease the other. Perhaps better, we should have a commitment to use the financial resources of the shared account in a way that represents the beliefs and desires of a fictitious joint entity. In most ordinary cases our beliefs are similar enough that we don't have to worry about that, and our desires don't conflict too much, so we can use our individual desires as proxies for the desires of the pair. (Most people don't find themselves sharing finances with someone whose desires are in deep conflict with their own, and when it happens, both usually try to get out of the situation.) But in making decisions that affect the operations of the household, or in betting on things that we know we disagree about, we need to be careful. We may in these sorts of cases agree to let the beliefs and desires of one of us win out for particular topics. Or we may have some compromise norm. But when acting with the joint account, we should do something of the sort, unless we have completely lost our commitment to work together.

I claim that something similar is at work diachronically. I have reason now to commit to certain plans for updating my beliefs in the future. This commitment may be as a result of accuracy considerations showing that this plan for update is the one that has the best expected accuracy over all possible futures. Or it may be because I see that I and my shared future will suffer a combined loss if I don't update in this way. However, if my future self has a radical change of heart, *he* will not be irrational to violate *my* current plan for updating, any more than *my partner* must be irrational if *he* has a change of heart and spends our joint account to run away to Chile. Both of these outcomes would be unfortunate for *me now*, and they would be bad for the hypothetical collective agent, but the one that makes the break has already decided to leave this collective agent behind. There is irrationality only if the subject involved intends to maintain the commitment to the collective while acting in this way. This idea is developed further by Easwaran and Stern (2026), and I will return to it in Sections 6.5.5 and 7.2.

6.4 ALTERNATIVE UPDATING RULES

Conditionalization is the standard Bayesian updating rule, and it is the one that is used in by far the most applications. However, we have seen that the argument for conditionalization requires some very specific assumptions. The subject must be in a situation where there is some partition of the possibilities into two, and they will have one posterior doxastic state in one set of possibilities, and another in the other. Most of the arguments have been given in a somewhat more general form, where the partition has a finite number of sets, rather than two. Rescorla (2018) and Nielsen (2021b) generalize the Dutch book argument to infinite partitions, while Easwaran (2013a), Nielsen (2022), and Easwaran and Nielsen (2025) generalize the Greaves and Wallace argument to infinite partitions. Both of these lead to updating by a "regular conditional probability", as described by Easwaran (2019).

However, there are some kinds of learning contexts that don't seem to fit these generalizations.

6.4.1 Learning without certainty—Jeffrey conditionalization

Richard Jeffrey (1965) noted that standard conditionalization always works by bringing some proposition to certainty, but there seem to be some learning situations in which no proposition is learned with certainty. For instance, you step into a fancy restaurant, dimly lit by candlelight, and look around—when you see the tablecloths, you're not sure what color they are, but based on what you see, you become somewhat more confident that they are blue or green, or possibly purple.

Jeffrey proposed a description of such updates, but not a justification of it. In the simplest version, Jeffrey imagines that there is some particular proposition e that drives the update, but that it doesn't go all the way to certainty. Instead, the agent's degree of belief in e goes from $P_0(e)$ up to some value $P_1(e) = x$, which could be quite high, but still short of certainty. In this case, Jeffrey suggests the experience shouldn't change the way that the evidence relates to any other proposition. That is, we should have $P_1(a|e) = P_0(a|e)$, and $P_1(a|\bar{e}) = P_0(a|\bar{e})$. Putting these together with the synchronic constraints on conditional probability, we see that $P_1(a) = P_1(a \cap e) + P_1(a \cap \bar{e}) = P_1(a|e)P_1(e) + P_1(a|\bar{e})P_1(\bar{e}) = xP_0(a|e) + (1 - x)P_0(a|\bar{e})$. In the special case where $x = 1$, we get $P_1(a) = P_0(a|e)$, so standard conditionalization is a special case of Jeffrey conditionalization.

Jeffrey further allowed that this could be generalized to a case where there is not just one proposition driving the update. If $e_1, \ldots, e_n$ are the n mutually exclusive and exhaustive alternatives that make up the evidential partition (perhaps these are the different colors that the tablecloth might be), then they can update by the rule $P_1(e_i) = x_i$, where the x_i add up to 1. Then other propositions update by the calculation $P_1(a) = x_1P_0(a|e_1) + \cdots + x_nP_0(a|e_n)$. The condition that $P(a|e_i)$ stays the same for any proposition a and any piece of evidence e_i is known as "rigidity", and it is what characterizes Jeffrey conditionalization.

Under Jeffrey's characterization, it is hard to say how to characterize the effect of the *same* experience on agents who had *different* starting degrees of belief. In ordinary conditionalization, we could just specify the proposition that functioned as evidence, but for Jeffrey, we need to specify both the evidence and the posterior of the evidence. But specifying the posterior directly leads to some odd features. If one subject starts at $P_0(e) = .1$ and goes to $P_1(e) = .9$, then it would be strange for someone else who started at $P_0(e) = .95$ and had the same experience to *decrease* her degree of belief in e. For this reason, this update is often referred to as "probability kinematics"—unlike a dynamical theory, it doesn't try to analyze the forces showing *why* things move, but just analyzes the kinematic connections of how other things *must* move, *if* these things move that way.

But Hartry Field (1978) and Carl Wagner (2002) and others have shown that it is better to characterize the update in a different way. Rather than directly specifying the value of $P_1(e)$, or even specifying the difference or ratio between $P_0(e)$ and $P_1(e)$, they suggest that we should specify the ratio between the prior *odds* $P_0(e)/P_0(\bar{e})$ and the posterior odds $P_1(e)/P_1(\bar{e})$. This "odds ratio" is also often called the "Bayes factor" $\frac{P_1(e)/P_1(\bar{e})}{P_0(e)/P_0(\bar{e})}$.[7] By characterizing the update this way, it turns out that we can apply multiple updates on different partitions in different orders and get the same end result.

These odds ratios or Bayes factors appeared in Section 6.2.6 when we were considering how a single piece of evidence affected different propositions in a partition for two subjects with the same likelihoods but different priors. Diaconis and Zabell (1982) show that this is no coincidence. Although there is no proposition in the subject's doxastic state that goes to certainty, we can always represent a Jeffrey update by adding one extra "virtual proposition" to the doxastic state, giving it a likelihood in each world and extending the probability function to it, and then doing standard Bayesian conditionalization on that proposition.

Skyrms (1987) has given arguments seeking to justify this kind of Jeffrey updating, both by Dutch book and by convergence to the truth.

6.4.2 Evidence and externalism

While Jeffrey's objection to standard conditionalization focused on the fact that it requires certainty in a proposition, another objection comes from the idea of externalism in epistemology. Historically, most philosophers sought to base a theory of knowledge on the kinds of things that the subject has access to. Descartes worried that much of our perception of the external world could be an illusion produced by an evil demon, so he tried to base all our knowledge on the kinds of reason that can be carried out inside the head, and (he thought) could provide certainty. But in more recent decades, many epistemologists have worried about this. They have noted that much of our knowledge depends on justifications that we just can't get access to. People throughout history have been justified in relying on their senses for beliefs about the world, even though it is only in the past century or two that we have had anything like a good explanation of how our senses manage to be reliable.

---

[7]In the case of a more complex evidential partition, we would specify each Bayes factor $\frac{P_1(e_i)/P_1(e_j)}{P_0(e_i)/P_0(e_j)}$.

One proposal, from Goldman (1979), is that we are justified in any beliefs that are produced by a reliable faculty, even if we have no reason to believe that this faculty is reliable. I'll return to something like this idea in Section 8.2, though the relevant concept of reliability will be relativized to a "home environment". Because reliability is a fact about the relation between the self and the outside world, and because two beings could be internal duplicates, but one have reliable faculties and the other not (consider an ordinary person and their internal duplicate in a Cartesian "evil demon" world), this sort of view is said to be "externalist".

One particularly influential externalist program in recent decades has been the "knowledge first" program of Williamson (2002). He takes knowledge as an un-analyzed concept, which he uses to analyze other epistemological concepts. In particular, he argues that one's "evidence" consists of the propositions that one knows. He argues that our perceptual faculties have "margins for error"—when I look at a 10 m tall tree, the strongest thing I know may be that it is between 9 and 11 m tall. But even if I know that my perception has this level of precision, I can't *know* that I know the tree is between 9 and 11 m tall—if the tree were actually 9 m tall, I would know that it's between 8 and 10 m tall, and if it were actually 11 m tall, I would know that it's between 10 and 12 m tall, and all I know is that I'm in one of these three situations. So the strongest thing that I *know* that I know is that the tree is between 8 and 12 m tall. The margin for error grows as we iterate the "know that" operation. Thus, many of the logical principles we want for knowledge must fail—a subject can know something without knowing that they know it; a subject can fail to know something without knowing that they fail to know it; not every truth is knowable (though Williamson accepts that everything that is known must be true). Also, there can be a certain kind of asymmetry of knowledge. When I am awake, I can know that I am not dreaming, but when I am dreaming, I don't know whether I am awake or dreaming. When I am looking at a red wall under normal light, I can know that I am, but when I am looking at a white wall under tricky red lighting, I don't know which of the possibilities I am in.

Das (2022) argues that this raises major problems for conditionalization. Under standard conditionalization, the possible pieces of evidence that one can learn form a partition. This means that in any world where the subject learns the evidence, it is true, and conversely, in any world where it is true, the subject learns it. Any world that is not ruled out by the subject's evidence is one in which the subject has the same evidence.

This means that conditionalization can't represent the kind of learning situation that we often find ourselves in with these tricky situations.

Williamson (2011) provides a standard example that many others have picked up, involving an unmarked clock. If the minute hand is actually pointing at :10, the lack of markings means that the most one knows is that it is pointing between :09 and :11. In this case, there are 60 possible worlds, and each one has a different evidence proposition, overlapping several others. The possible evidence propositions do not form a partition, as was required in all the arguments for conditionalization.

Schoenfield (2017) argues that in this case, conditionalization is in fact *not* the optimal update rule in this situation. She instead shows that the optimal update rule is the one sometimes called "auto-epistemic updating" (Hild, 1998)—instead of conditionalizing on the evidence, condition on the proposition *that* it is the evidence. This certainly yields better results—in fact, it yields certainty that the hand is pointing at :10, even though the subject is not in a position to know that. Various other authors have provided alternative analyses of this case based on update rules that make use of the evidence, though in some cases, one updates on the *wrong* evidence by accident (Carr, 2021, Gallow, 2021, and Schultheis, 2023).

But I think that relying on the evidence proposition is somewhat of a distraction. Isaacs and Russell (2023) suggest a proposal in which evidence does not play a role—instead, in each world, the subject has some probability of "guessing" that they are in any one of a number of worlds (the ones that Williamson and others describe as being part of the evidence proposition). They show that the optimal update here is a kind of Jeffrey update that shifts each world's probability in proportion to the likelihood it gave to guessing that one was in the world one guessed.

But I think the use of the guess about the world is also inessential. Easwaran and Nielsen (2025) argue that the right way to think about these situations doesn't involve a proposition at all. Rather, there is just some feature of the world that the subject becomes sensitive to as part of their update, and as a result they receive some "signal" that they can use to try to improve their accuracy or their decisions. If this signal is a deterministic function of the world, then we get the standard scenario that justifies conditionalization. But not when the signal is indeterministic. If each position of the clock's hand has some likelihood of yielding a given visual experience, then on receiving that visual experience, each world's probability shifts in proportion to that likelihood. This is much like the idea of Isaacs and Russell (2023), but with the possibility that there might be no particular correspondence between signals and worlds—there might be more possible signals than possible worlds. (This is very much like the idea of Konek (2023).)

Much philosophical discussion of epistemology has assumed that there is some special role played by a proposition that we can call the "evidence". In the kinds of simple cases that give rise to conditionalization, this is a useful way to talk—the update happens in some worlds and not in others; we can identify some proposition e with this set of worlds; and we can calculate the probabilities of all propositions in the best update by intersecting them with e and dividing by the probability of e. It is thus harmless to think that somehow the proposition e is driving the update. But in other cases, like the unmarked clock or the trickily lit wall, it's a mistake to focus on a proposition as driving the update. The subject undergoes some experience, and gains some abilities to make their doxastic state a bit more sensitive to how the world is. What matters is that they use these abilities as effectively as they can for whatever doxastic goals they have, whether pragmatic or alethic. Epistemologists who want to say that a proposition plays the role need to say something about how the learning event gets this propositional content, and what it means that the subject sometimes updates on the "wrong" content. But if we think more directly in terms of abilities than in terms of propositions, it is more straightforward.

6.4.3 Maximum entropy

One other update rule has been discussed by Jaynes (2003). He suggests that we can measure the "distance" between two probability functions by computing the "entropy" of one with regards to the other. It turns out that this entropy from P to Q can always be computed by taking the difference in expected accuracy that one probability function gives to the other, for some proper scoring rule (usually the logarithmic rule). In this context, entropy is thought of as showing the amount of information needed to get from one distribution to the other.

Jaynes imagines an update being driven by some constraint on the posterior probability, and then recommends the update rule that moves to the probability distribution that maximizes entropy out of the ones that satisfy the constraint. This is the one that is in some sense closest to the prior. Skyrms (1985) shows that there are certain conditions under which updating by maximizing entropy agrees with conditionalization, or Jeffrey conditionalization, and argues that this update rule can be justified in these situations but not others. Paris (1998), Landes and Williamson (2013), and Lukits (2014) argue that maximizing entropy gives intuitive results that agree with common sense, and provides a unique objective way to drive updates in situations not covered by the other rules. But none of these justifications follows directly from the pragmatic or alethic role for the doxastic state, so I am less convinced that it is doxastically required.

6.4.4 New possibilities

The arguments for conditionalization involved situations where the subject gains a new sensitivity, and is able to rule out some possibilities. While the motivations for Jeffrey conditionalization, externalist evidence, and maximum entropy involved other kinds of learning situations, they still generally either involved the subject eliminating doxastic possibilities, or somehow restructuring their doxastic state over the same possibilities. But at least sometimes, subjects *increase* their space of doxastic possibilities. One way this happens is in cases of forgetting—the subject initially had some possibilities, then learned something that ruled it out, and then forgot what they learned so the possibility came back. But another quite different way it happens is when the agent realizes there are certain possibilities they had just never even considered. The detective had only been considering the guests at the hotel as potential suspects, and then realizes that they should have been considering the staff, and residents of buildings near the hotel.

It is very difficult to provide any forward-looking guidance on how to do this. If there are possibilities that the subject isn't considering, then it's hard for them to have any plan for how to consider them. In cases of forgetting, where the truth really was among the possibilities they were considering at the earlier time, but now the subject returns to considering a wider range of possibilities, the Dutch book and accuracy dominance arguments for conditionalization still work just fine—the subject will do better overall if their degrees of belief at the time when they have more information is related by conditionalization to their degrees of belief at the time they have less information, even though these two times occur in the opposite order. This and related ideas are discussed extensively in Titelbaum (2012). But in the case where the subject just realizes there are possibilities they forgot to include, with no good reason, it's unclear that anything can be said. There are attempts to justify particular rules, like that of Maher (1995), but this strikes me as an attempt to say what is rational to think when someone is not thinking rationally. I don't think it makes sense to give a systematic answer, any more than it does for setting priors. Degrees of belief will change, but this change can't be subject to rational rules.

6.5 REFLECTION

One other controversial principle that is related to the update of probabilities is the principle of reflection. This principle was introduced by van Fraassen (1984), as a principle entailed by Bayesianism that he thought was obviously *false*, but was then discussed further in his (1995), by which point he thought it was obviously a principle that rational doxastic subjects should at least be *committed* to. There are several formulations of principles that go by this name, ranging from mere observations of basic mathematical facts, to substantive principles of diachronic rationality. What they have in common is that they express some way that one's current degrees of belief should reflect one's best estimates of one's future degrees of belief.

6.5.1 The Law of Total Probability

The first observation in this direction is a simple mathematical theorem that has already been referred to several times. If $e_1, \dots e_n$ form a partition, then a just is the proposition $(a \cap e_1) \cup (a \cap e_2) \cup \cdots \cup (a \cap e_n)$. Since the e_i are incompatible, this means $P(a) = \sum_{i=1}^{n} P(a \cap e_i)$. By the definition of conditional probability, $P(a \cap e_i) = P(a|e_i)P(e_i)$. Thus, we have:

> Law of Total Probability: When $e_1, \dots, e_n$ form a partition, $P(a) = \sum_{i=1}^{n}(P(a|e_i)P(e_i))$.

This fact was used in footnote 3 to fill in the denominator of Bayes's Theorem from priors and likelihoods, and was used to derive Jeffrey Conditionalization by assuming that conditional probabilities are fixed while the priors of a partition change to new posteriors. Several of the arguments for conditionalization also made use of this fact in a slightly different form. If P_{e_i} is the result of conditionalizing P on the proposition e_i, then the theorem states that for all a, $P(a) = \sum_{i=1}^{n} P(e_i)P_{e_i}(a)$, that is, that P is a "mixture" of the probability functions P_{e_i}.

If a subject is certain that they will update by conditionalization on this partition, then they have degree of belief $P(e_i)$ that they will conditionalize on e_i, and so their current expected value for their posterior in a is $\sum_{i=1}^{n} P(e_i)P_{e_i}(a)$. This yields a version of the reflection principle—one's current degrees of belief should equal one's current expectation of one's updated degrees of belief.

6.5.2 No certainty of confirmation

Among other things, this implies that one's current degree of belief in any proposition should be somewhere in between the highest and lowest degree of belief that one might have in that same hypothesis after conditionalization. Put another way, for any update, there should be no proposition such that it is certain that the update will result in an increased degree of belief in that proposition (and similarly for certainty of decrease).

Put like this, the principle is eminently plausible, no update should promise a guarantee of confirmation in advance. If one were certain that the update would result in confirmation, one should already have had a higher degree of belief in that hypothesis. So far, we have derived it as a mathematical conclusion from the assumption that update proceeds in line with conditionalization on a finite partition. But one can give a Dutch book for this more directly. If there is some $x < y$, and someone currently has degree of belief below x in a, but is certain that at some specific future time they will have degree of belief above y in a, then they will currently sell a bet on a for x, and are sure that they will be willing to buy it back for y, and guarantee a loss of at least $y - x$. The more general accuracy-based argument for this principle is less straightforward, and is given in Easwaran (2013a).

The version of reflection stating that one's current degree of belief should be somewhere in the range spanned by one's possible future degrees of belief is plausible enough that van Fraassen (1999) takes it, without argument, as the basis for a new argument in favor of conditionalization. Additionally, many philosophers have tried to generalize a version of this principle to other types of attitude, though there are questions about how precisely to formulate it for utilities or imprecise probabilities (Moss, 2020).

But as we will see shortly, there are also important challenges even for the straightforward formulation for degrees of belief.

6.5.3 Similar principles

If the subject currently has doxastic state P_0, and is certain that they will update by conditionalization to reach P_1, then we can reformulate this principle yet again. Let "$P_1(a) = x$" be a name for the proposition that is true in all the worlds where the subject will update to a P_1 such that $P_1(a) = x$. We can then show that $P_0(a|P_1(a) = x) = x$. Because this formulation involves statements of probability *within* the probability function, it doesn't directly follow from principles discussed so far, which are all content neutral.

Various other principles discussed over the years have given formally similar principles with different probability functions in the condition. If we replace "P_1" with a function representing the observed frequencies of a within a population, then we get Pollock (1983)'s principle of "direct inference". If we replace "P_1" with a function representing the objective chances of a, we get Lewis (1980)'s "Principal Principle". If we replace "P_1" with a function representing the degrees of belief of some expert, then we get Elga (2007)'s principle of "expert deference".

Gallow (2017) shows that it is impossible to satisfy multiple principles of this form unless one is certain that the relevant probability functions will agree. Pettigrew and Titelbaum (2014) deals with some important subtleties around the formulation of these principles.

None of these principles are requirements of Bayesianism. There have been various arguments given for these principles, but I think that they each build in a controversial normative role for observed frequencies, objective chances, or the doxastic states of experts. I instead think that the version for observed frequencies should be rejected, except in the case where one currently has uniform degrees of belief about what frequencies one will observe (in which case it is a theorem). I follow de Finetti (1974) in reversing the interpretation of the Principal Principle—rather than saying that *if* one takes a probability function to be the objective chances, *then* one's degrees of belief in any proposition should be one's expectation of the value of the probability function, I say *if* one's degrees of belief in any proposition are one's expectation of the value of the probability function *then* the probability function can be treated as the "objective chances". Similarly, rather than saying that *if* one treats someone as an expert, *then* one's degrees of belief should be one's expectation of their degrees of belief, I say that *if* one's degrees of belief in a proposition are one's expectation of a person's degrees of belief in that proposition, *then* one is treating that person as an expert.

6.5.4 Challenges to reflection

Many challenges have been raised to versions of the reflection principle. Some classic challenges come from Christensen (1991) and Talbott (1991). For example, if I know that I am about to drink alcohol or take a drug, I may be able to foresee that my degrees of belief will change in a particular direction. I currently have low degree of belief that I can safely drive after three drinks, and while I'm not certain how confident I will end up after those three drinks, my current degree of belief is probably less than the weighted average of those possibilities, and perhaps even below the lowest value it might end up. My degree of belief in my driving ability after three drinks, conditional on my future self being highly confident in that ability, is not high. This is the sort of case suggested by van Fraassen (1995)'s reference to "Ulysses and the Sirens".

Another sort of example comes from foreseen memory loss. I am currently confident that I went for a bike ride on Thursday evening. But I am highly confident that, a year from now, my degree of belief that Thursday was the evening I went on a bike ride will be lower. My current degree of belief is quite a bit higher than my current expectation of my future degree of belief. My current degree of belief that the bike ride was Thursday, conditional on my future self having degree of belief 1/7, is just as high as my unconditional degree of belief, not 1/7.

These two challenges both occur because I am currently confident that my future degrees of belief will be sub-optimal, by my current standards. My violation of reflection indicates a kind of irrationality, but it is not that I am *currently* irrational for violating it now—I am confident that my *future* self will be irrational. These cases suggest that reflection is better interpreted as a wide-scope norm (Broome, 2007) than a narrow scope norm—it ought to be the case that a subject's current degree of belief is the expectation of the subject's possible future degrees of belief, rather than the expectation of the subject's possible future degrees of belief being the degree of belief the subject currently ought to have.

A more subtle counterexample comes from Frank Arntzenius (2003). I present a somewhat less science-fictional version of this scenario here than the "Shangri-La" problem Arntzenius describes. Even if I know I won't undergo any forgetting, my future self might think that I have. Loftus and Pickrell (1995) have shown that it is possible to get people to develop false memories by asking them leading questions about a partially-remembered event. A psychologist might describe the following setup: first, a coin will be flipped and I will see how it comes up; a few weeks later, if it came up heads, they will prime me to develop a false memory that it came up tails; then several weeks later, they will remind me of the whole setup. At the end of the process, I will have either a real or apparent memory of the coin coming up tails. Since I am aware of the process, and my future self at the end of the process will be too, I expect that future self will have credence .5 that it did in fact come up tails.

Right now, while I am looking at the coin having come up tails, I am quite confident (if the psychologist is trusted about the setup) that the psychologists will not tamper with my memory. However, I know that my future self at the end of the process will worry that they did, and will thus only have credence .5 that the coin actually came up tails. And yet, I currently have very high credence that the coin came up tails, and so I violate reflection.

This time, one can't blame the failure of reflection on the future degrees of belief. They seem like a perfectly reasonable response to a situation that *could* have been unreasonable, for all the future self can tell. But still, someone who was never prone to forgetting or false memory, and was certain of this fact about themself, would not have this violation.

Ray Briggs (2009) notes one other kind of challenge. If, as I suggest in Chapter 8, there are multiple credal states that are all rationally compatible with the same evidential situation, then one might change from one to another in a predictable but non-rational way. This might occur due to what Laurie Paul (2014a) calls a "transformative experience", such as becoming a parent, joining a new community, undergoing a religious experience, or the like.

Someone who is about to have a child may be considering whether or not certain early childhood interventions are important—for instance, whether speaking to children in a childish way rather than as if they were a small adult has a significant impact (positive or negative) on the rate of vocabulary acquisition. Current research on topics like this is highly equivocal. But I suspect that most parents tend to believe this sort of thing is highly important, while most non-parents do not, and both may be rationally responding to the equivocal evidence we have. A person who is about to have a child may currently be highly confident that this intervention is *unimportant*, but also be highly confident that, although they are somewhat uncertain about precisely how the experience will change them, they will eventually be highly confident in the *importance* of these interventions. This would be another violation of reflection that seems like it might occur for a person who is rational at each moment, in a case where rational people might disagree about how confident to be in the importance of these interventions.

Sarah Moss (2015b) and Brian Hedden (2015) argue that again, this violation of reflection is an indication that something irrational or unreasonable is going on. Both of them defend what they call "time-slice epistemology", according to which rationality is always synchronic, and a violation of an apparent diachronic norm like conditionalization or reflection is just an indication that at least one of the two time-slices was synchronically irrational. However, in this sort of case (and many others), to defend this view, one must defend the idea that for any given body of evidence, there is a unique set of degrees of belief that would be rational to hold. For more on this debate on uniqueness and permissivism about rationality, see Horowitz (2013), Horowitz and Dogramaci (2016), and Kopec and Titelbaum (2016), among others. (I will return to this issue in Section 8.1.)

Ray Briggs (2009) develops a weaker formulation of reflection that they call "distorted reflection". Rather than requiring one's current degree of belief in something to *equal* the average of the possible values one thinks it might have in the future, it only requires that it be *close*, with a distorting factor that takes into account the extent to which one's current or future self might be under the influence of irrationality, or a different epistemic standard.

6.5.5 Reflection as commitment

But I think that an important aspect of these scenarios is already addressed by van Fraassen (1995). He notes that although reflection is formulated in terms of a condition on one's current degrees of belief (it must be the average of one's possible future degrees of belief), it is better understood as an expression of one's current *commitments* to how one's future degrees of belief *ought to be*. Although one might recognize that one's future self *will* have degrees of belief other than the ones one would end up with through conditionalizing (either through drugs, or forgetting, or worries about having had one's memory tampered with), one's current *commitment* ought to be to ones that would conditionalize.

This is similar to what I said about similar principles in Section 6.5.3. It's not that a subject *should* defer to people who are antecedently described as experts—it's that having this kind of deference to them is what *makes* them an "expert" according to the subject.

Marušić (2015) says that one should believe in accordance with these commitments. Briggs modifies reflection to account for ways one's future degrees of belief might diverge from one's commitments. Reflection itself just states the consequences of believing in line with those commitments. Dorst et al. (2021) shows how to modify many of these principles to account for the possibility of some "expert" being uncertain about living up to their own commitments, and derives these principles from a commitment by the subject to trust that the expert will make better decisions according to the subject's own values (whether pragmatic or alethic).

6.6 CONCLUSION

In this chapter I considered the diachronic question of how a subject's doxastic state should change over time, and the related quasi-diachronic question of how the subject should currently *plan* to change their state, or be *committed* to changing their state. I gave the mathematical definition of conditional probability as $P(a|b) = \frac{P(a \cap b)}{P(b)}$, and described how it could be applied in conditionalization.

I showed that update by conditionalization gave rise to a useful Bayesian theory of confirmation, which reflects certain intuitive principles about entailment and induction, but slightly revises some other principles about induction, and confirmation by instances. There is no systematic syntactic theory of confirmation, but Bayes' Theorem gives the closest well-justified approximation, which makes use of priors and likelihoods. However, there is a problem of the priors—instances of confirmation that seemed to be objective turn out to depend on a subject's priors, even when different subjects agree about the likelihoods.

I then showed how conditionalization can be justified from both the pragmatic and alethic role for the doxastic state, in situations where the subject acquires a capacity to change their doxastic state in a deterministic way across a partition of worlds. There were dominance-based arguments for a truly diachronic norm, and expectation-based arguments for a quasi-diachronic norm of planning to conditionalize. But there were worries about the extent to which any truly diachronic norm can be rationally binding on the subject—though it is more plausible that the quasi-diachronic norms can be justified as expressions of a commitment.

I discussed several alternatives to conditionalization that have been proposed for more general learning situations. While some philosophers think that there must be an evidence proposition that plays a special role in determining the rational update, I think that what is more important from the subject's perspective is just that they use whatever abilities they have to the best extent possible. When these abilities can be summarized by a proposition, we get conditionalization, but in other cases, the best they can do will look more like a Jeffrey update.

I then considered versions of the "reflection principle", and related it to the Law of Total Probability. I showed that if a subject is certain they will update by conditionalization, then the reflection principle follows, but that otherwise versions of it might be questionable. I compared it to related principles of deference to frequencies, chances, and experts. I argued that all of these principles are best understood as expressions of commitments, rather than as rational requirements.

RESEARCH QUESTIONS

1. How do the actual learning situations faced by actual humans, animals, groups, and other agents relate to the kinds of idealized learning situations for which conditionalization is defined?

2. What is an adequate account of confirmation?

3. How do we deal with the problem of the priors?

CHAPTER 7

Three components of rationality

In Chapter 1, I argued that there is a general conception of a doxastic state for subjects that have either a goal to represent the world for either pragmatic or alethic purposes. Such subjects include humans, but also many other types of creatures, possibly including animals, artificial intelligences, associations, aliens, and angels. In Chapters 2 and 3, I developed versions of this concept on which the primary role is pragmatic (guiding action) or alethic (aiming at the truth) and showed that either way, the doxastic state must satisfy probabilism in order to effectively play the relevant role. In Chapter 4, I argued that the numerical nature of probabilism does not make it as difficult a condition to satisfy as some philosophers have thought. Many apparently different formalisms are all equally good ways to represent probabilistic degrees of belief, and arguments that actual humans can't obey probabilism ignore the other equally difficult representational tasks our minds manage to do. Worries about degrees of belief being imprecise often arise because our degrees of belief are not stable or, as considered in Chapter 5, because we make mistakes about what it means to have the same numerical degree of belief. In Chapter 6, I considered the concept of conditional degree of belief as an explication of the way an agent ought to plan to update over time.

In this chapter, I consider the force of the various rational requirements that an agent is under. In particular, I suggest that there are actually three separate sorts of normative requirement. In particular, the synchronic norms of probabilism and logical omniscience, and the diachronic norm of conditionalization, have three separate sources. Thus, to the extent that violation of one of the three is a type of irrationality, they are different types of irrationality. Towards the end of the chapter, I consider cases where someone is said to be irrational despite satisfying these three norms. In the next (and final) chapter, I will suggest that this is yet another conception of rationality at work.

7.1 PROBABILISM AND LOGICAL OMNISCIENCE

Chapter 2 gave pragmatic arguments for probabilism. Degree of belief is taken to be constitutively characterized as a mental state that governs preference among actions, directed at one's pragmatic values. In a simplistic version, the actions are bets, the degrees of belief correspond to the ratio of price to payoff, and anyone who violates probabilism will accept a set of bets individually that guarantees a loss. In the more sophisticated version, the actions are any behaviors whose outcome depends on uncertain features of the external world, the payoffs and costs are whatever the agent actually cares about, and the measurement comes from structural features that allow us to create a scale—and again, if the degrees of belief don't satisfy the probability axioms, then there are actions the agent prefers that collectively guarantee that things go badly according to her values. If the characteristic role of degrees of belief is to govern behavior to bring about things the agent values, then these degrees of belief are counterproductive if they aren't probabilistic.

Chapter 3 gave alethic arguments for probabilism. Degree of belief is taken to be constitutively characterized as a mental state that aims at overall closeness to the truth, measured by some "proper scoring rule". In Joyce's (1998, 2009) version of the argument we put some specific constraints on the scoring rule, and then show that if one's degrees of belief fail to satisfy probabilism, then there is another set of degrees of belief that is guaranteed to be closer to the truth, regardless of what the truth is. Predd et al. (2009) show that this argument works with any proper scoring rule. Lindley (1982) and Ajalloeian (2025) show that the requirement that it is a proper scoring rule is just an artifact of the choice of representation—as long as degrees of belief are represented numerically, and inaccuracy is measured in some differentiable way, there is a transformation of the scale that turns this measure into a proper scoring rule.

7.1.1 Guarantees

Both of these types of arguments take the form of starting with a normative value (either the agent's practical desires, or closeness to the truth) and then showing that unless one's degrees of belief conform to probabilism, one is guaranteed to do badly according to this value. Thus, both arguments seem to suggest that a violation of probabilism is irrational, since a violation of probabilism guarantees bad results according to one's own values. However, they leave open the question of what "guarantee" means.

Consider someone who has non-extreme degree of belief in some proposition. If the proposition is false, they would have been better off if they had maximum confidence in the proposition—pragmatically, this is because a degree of belief less than maximal would lead them to turn down some bets or acts that would actually have turned out well for them, and alethically, this is because a degree of confidence less than maximal is farther from the truth than maximal confidence. Thus, an intermediate degree of belief is, in some sense, guaranteed to be worse than an extreme one. However, no one thinks that it is irrational to fail to have an extreme degree of belief.

The standard reason is that although it is guaranteed that some extreme degree of belief is better, in the possibilities where the proposition is true, the highest extreme degree of belief would be better and the lowest would have been worse, while in the possibilities where the proposition is false, the reverse is true. There is no particular degree of belief that is guaranteed to be better than the intermediate degree one has. So there is no practical or alethic self-defeating aspect to having this intermediate degree of belief. Omniscience is not required for rationality.

However, Saul Kripke (1972) argues that there are some necessary truths that we can't know a priori. For instance, if Hesperus (the evening star) and Phosphorus (the morning star) are in fact the same object, then it is impossible for them to have been different objects. So if someone has intermediate credence that Hesperus is Phosphorus, then it is impossible that their credence function guide them as well as a credence function that assigned certainty to the proposition that Hesperus is Phosphorus. And yet, we don't want to claim that someone is irrational for being uncertain of this proposition.

Similarly, consider the proposition that 10,537 is a prime number. I have verified that it is not divisible by 2, 3, or 5, but I'm not sure whether it's prime. I have middling credence, probably around 0.3 on the probability scale from 0 to 1, that it is prime. And yet, if it is in fact prime, then it is impossible for my credences to be as accurate as assigning maximal credence to this proposition, and if it is not prime, then it is impossible for my credences to be as accurate as assigning minimal credence to this proposition. In this case, there are even some propositions I do believe that logically entail the correct answer to this claim. I am sure that addition and multiplication are associative and commutative, that multiplication distributes over addition, and that 10,537 is the sum of 10,000, 500, 30, and 7. These propositions and a few others, together logically entail every fact about multiplication and primality, and so must either guarantee that 10,537 is prime, or guarantee that it is not.

Any being that is not "logically omniscient" is thus guaranteed to do worse than one that is. And yet logical omniscience (having certainty in all logical truths, and minimal credence in all contradictions, and having higher credence in any proposition entailed by other propositions one has credence in) seems like an overly strong requirement, for humans or communities or even artificial intelligences. In fact, results by Gödel and Turing show that there is no recursively computable function that assigns 1 to all logical truths and 0 to all contradictions. So satisfying logical omniscience would require violating the version of the Church-Turing thesis that says that every function realized by a physical system is recursively computable!

However, it seems to me that the response we made to the claim that rationality requires actual omniscience will also defuse the idea that rationality requires metaphysical or logical omniscience. Any agent that has intermediate credence in a logical or metaphysical truth would be better off if they had certainty in this proposition, whether because they would perform some actions that leave them better off (for the pragmatic argument) or because their credences would be more accurate (for the alethic argument). However, as long as this agent thinks there is some possibility of the proposition being true, and some possibility of it being false, they have no way to guarantee that certainty is the better option, rather than certainty in its falsehood.

7.1.2 Possibilities

Following Hacking (1967b), and as I suggested in Section 1.3.7, I claim that it is the agent's own sense of possibility that we should use to understand the concept of "guarantee" that appears in these arguments.

For someone's preferences among acts to aim at some personal conception of value, they must not prefer act A to act B if they don't find it possible for A to result in a more valuable outcome than B. For someone's degrees of belief to aim at accuracy, they must not have credences C if they don't find it possible for C to be more accurate than D. They aren't really "aiming" from their own perspective if they don't satisfy these conditions.

However, if an agent fails to be logically omniscient, this means that there is a logical truth T such that they find it possible for T to be true, and also find it possible for T to be false. There are propositions A and B such that A logically entails B, but the person finds it possible for A to be true while B is not. Failures of logical omniscience mean that if we represent a proposition with the set of "doxastic possibilities" in which it is true, then some logical truths (and metaphysical truths, and actual truths) will be represented by sets that fail to include all doxastic possibilities. There will be propositions A and B such that A logically entails B, and yet the set of doxastic possibilities corresponding to A is not a subset of the set corresponding to B. (See Skipper and Bjerring (2019) for more discussion of this idea.)

An agent can satisfy probabilism in the form of the probability axioms over such a broader set of doxastic possibilities, without satisfying logical omniscience. All that matters is that for each proposition, the credence is non-negative; for a proposition that includes all doxastic possibilities, the credence is 1; and if a and b are propositions corresponding to disjoint sets of doxastic possibilities, then the credence in the proposition corresponding to their union is the sum of the credences in these two propositions. But "including all doxastic possibilities" doesn't necessarily correspond to logical necessity, being disjoint doesn't necessarily correspond to logical incompatibility, and the union of two propositions doesn't necessarily correspond to logical disjunction.

Note that there are actually two kinds of ways that logical omniscience can fail. I have so far focused on the kind of failing in which one finds it possible that a logical or mathematical truth be false. But there are also failings in which one fails to find something possible. If we consider what credence Newton had in the relativity of simultaneity, or the scientific community as a whole in the 17th century, it seems likely that they didn't even conceive of this as a possibility. (Few if any people did before Einstein.) They treated the absoluteness of time as a certainty, even though there are various mathematical truths that they treated as uncertain. In this case, it appears that this certainty was false—the actual world doesn't correspond to *any* of these doxastic possibilities. Something like this is likely true for us as well. The set of doxastic possibilities includes things that are not actually possible, and it might fail to include things that are in fact possible, or even actual. As I mentioned in Section 6.4.4, I'm not sure that there is anything to say about how to rationally incorporate new possibilities—the relevant kind of rationality will just say that one should have incorporated these possibilities already. See Chapter 5 of Earman (1992) for more on these two conceptions of logical omniscience, and their connections to challenges to conditionalization.

One might think that the conception of propositions as sets still builds in a kind of logical omniscience. The complement of the union of two sets is the intersection of their complements. And yet, just as someone can be ignorant of De Morgan's laws for propositional logic, one could be ignorant of De Morgan's laws for set theory. So it appears that formulating the laws of credence at the level of sets builds in a kind of omniscience.

However, a Bayesian agent doesn't generally explicitly reason with these sets of doxastic possibilities. When reasoning is explicit, it tends to occur at something much more like a linguistic level. When explicitly considering the sets, they might think there is some possibility of something being in the complement of the union of two sets without being in the intersection of the complements. But this just means "being in the complement of the union of two sets" and "being in the intersection of the complements of the two sets" are two propositions that they represent with different sets of doxastic possibilities! If there are two propositions that the agent represents with the *same* set of doxastic possibilities (which might well be the union of the complements of two sets, and thus the complement of the intersection of those same two sets), then they really don't find it possible for one proposition to be true without the other. These sets are a representational tool that the agent might use without having a full conception of the sets, just as a person can make use of gravity without having a conception of gravity. I claim it is best to think of the requirements of credal rationality as operating at the level of these sets of possibilities, rather than at the linguistic level. An agent that violates probabilism prefers one option to another, even though by their own sense of possibility and their own conception of value, they see no possibility for this first option to be better than the second. This is a fully internal conception of rationality that doesn't give rise to any requirement of logical omniscience.

This is a sort of requirement that Broome (2007) calls a "narrow scope" norm. *If* the agent has a given representation of propositions with sets of doxastic possibilities, *then* the agent is in fact rationally required (by the very nature of *aiming* at truth or practical value) to have credences over these propositions that accord with probabilism in its set formulation.

7.1.3 The importance of logical omniscience

However, there is still something to be said for logical omniscience. Unlike omniscience about necessity and omniscience about truth, there is something plausible about the claim that rationality requires one to believe the conclusions of logically valid arguments. When we see an argument and take it as valid we often *do* eliminate the possibilities where the premises are true and the conclusion is false. And there may be some thought that understanding of the language alone is all it takes to see these consequences, so every subject *can* perform these inferences.

My central claim in this section is just that if this is true, then the nature of that rational requirement is quite different from the rational requirement of probabilism. A violation of probabilism over one's own doxastic possibilities is a failing *according to* one's own conception of possibility. However, a violation of logical omniscience, if it is a failing, is a failing *of* this conception of possibility. There are various arguments for why logic should be normative on thought (MacFarlane, 2004 and Steinberger, 2017, 2019), but to the extent these arguments succeed, they will provide rational requirements on doxastic possibility. Rational requirements on confidence flow from doxastic possibility, so logic will be indirectly normative on confidence. To the extent that logical omniscience (or, as Chalmers (2012) claims, "a priori scrutability", or any other general requirement) is a norm, it comes from a very different source than probabilism, which comes from the very idea that degree of belief either guides practical actions towards one's values, or aims at the truth.

As I will suggest in Section 8.2, a disposition to accept these logical certainties will serve every subject well. But as I will suggest in Section 8.5, there may still be more pressing improvements that a subject can make to their doxastic state.

7.2 DIACHRONIC NORMS

In the previous chapter, I discussed the diachronic norm of conditionalization, and some alternatives that have been proposed. In Section 6.3.6, I discussed an objection raised by David Christensen to the effect that it is hard to see how to properly justify truly diachronic rules, since many of the justifications look relevantly similar to justifications of obviously silly rules, like a rule saying that one's degrees of belief should *never* change, or that a person and their spouse should always agree on everything. The idea is that whatever relation degrees of belief at different times bear to one another, it's not one that requires them to cohere with one another in the same way that degrees of belief of the same person at the same time should.

In Section 6.5.5 I retreated to the idea (partly attributed to Bas van Fraassen) that agents should at least *plan* to update their degrees of belief in line with conditionalization, and that the future degrees of belief they are *committed* to at a given time should satisfy reflection. But the question is whether this results in any truly *diachronic* norm.

The first thing to note when discussing diachronic norms is that the question of who they are even supposed to apply to can sometimes be somewhat tricky. The question of what it takes to be the same person over time is a notoriously difficult one. In ordinary cases, it's clear how a human body persists through time (through growth and development and aging) and it's clear that the person constituted by that body is the same person. But in some cases it's less clear. A person who undergoes a radical conversion experience and disavows the earlier self may be said to not be who she once was. This may be even more plausible in the case of severe amnesia followed by the development of new character traits.

Some even more unsettling cases emerged in the middle of the 20th century when one experimental treatment for extreme cases of epilepsy involved severing the connection between the two hemispheres of the brain (Nagel, 1971). Subsequent studies of these patients indicated that although they functioned normally in most circumstances, in situations where the two halves of the visual field were obstructed from each other, and where the two hands were presented with different stimuli, more unusual behavior was possible. Because speech is primarily controlled by one half of the brain, patients were only able to verbally identify objects presented visually in one half of the visual field. However, when objects were presented visually to the other half of the visual field, the patient would verbally deny that they saw anything, but would nevertheless successfully point to another object of the same type with the relevant hand. In ordinary life, the two hands and the two eyes appear to function together as well as they do in anyone else, but in the special setting where their stimuli are separated, they seem to behave like two separate people.

In his discussion of these cases, Nagel suggests that it is unsatisfactory either to describe the patients as having one mind or having two. Instead, he suggests that we shouldn't characterize mental phenomena in terms of minds, but instead just accept that various mental phenomena can be integrated to a greater or lesser extent. Humans in ordinary conditions don't exactly have true unity of consciousness, but instead have a very great degree of integration, produced by interactions through shared sensory perceptions, direct neural connections between the hemispheres of the brain, shared contact through the spinal cord to the lower parts of the body, and many other pathways. When enough of these pathways are interrupted, mental phenomena can organize in ways that seem more like two persons in one body. On this view, the notion of "one person" seems to be an exaggeration, even when just considering the mental phenomena at one time, let alone across time, as the split-brain patient moves between the greater integration of ordinary life and the greater dissociation of the lab setting.

A study by Nina Strohminger and Shaun Nichols (2015) of the effect of neurodegerative diseases on personal identity suggests future directions for some of this work. Caregivers (usually the spouse or adult child) of people with Alzheimer's disease, ALS, and frontotemporal dementia were asked about the severity of various physical or cognitive symptoms in the patient, and were also asked how much the patient "is still the same person underneath" or "seems like a stranger" and whether the caregiver feels like they "still know who the patient is". What they found is that in this third-person assessment, the symptoms that were most strongly associated with changes in who the patient is are changes in moral traits, more than loss of memory or speech or non-moral personality changes. Great care is obviously needed in assessing the relevance of this sort of work, but it suggests another avenue of research to figure out whether and to what extent the persistence of commitments (of values, beliefs, or other kinds) can be said to be constitutive of identity rather than guided by it.

If we are considering Bayesianism as a theory of a group, like the scientific community or a business, these sorts of worries are even more pressing. Why should a group composed of different people at some later time be in any way beholden to the plans of the earlier group? And even for an artificial intelligence, the question arises. One could design such an intelligence in a way that forces it to update in coherent ways, but is there a reason one ought to, or that the intelligence ought to keep doing so, if it finds some way to change its own design?

7.2.1 Parfit and impersonal coherence

In his book *Reasons and Persons* (1984), Derek Parfit supplements the empirical cases of human dissociation with various science fictional examples involving teletransporters and duplicators that create physical bodies bearing greater or lesser degrees of physical and psychological continuity to earlier ones. Motivated by these cases, Parfit suggests that what we ordinarily think of as a precise relation of being the same person should be replaced by various relations of continuity that come in greater or lesser degrees. We might think that cases of attenuated "identity" call for attenuated prudential norms of self-interest. But Parfit suggests that it might be better to replace these prudential norms of self-interest with more impersonal moral norms, and that the reasons we have for action should treat our future selves no differently from anyone else.

On a Parfitian view, we might say that the attenuated degree of continuity constituted by particular degrees of physical and psychological continuity might give one an attenuated reason for coherence across time. Over longer periods of time the reasons for continuity might be replaced by reasons for interpersonal coherence, if there are any. Some of the literature on disagreement and testimony has suggested that there are—mere awareness that your opinion is different from that of another person who is in no appreciable way epistemically worse off than you may give you a reason to modify your degree of belief to be closer to that of the other person. This view is controversial, but it may provide a natural alternative for grounding truly normative diachronic rules for Bayesianism, by introducing them in tandem with interpersonal norms. This would be a way to accept Christensen's analogy of the diachronic and interpersonal cases, but to draw the opposite conclusion. Not only does a person have reason to cohere with their future self, but a person and their spouse ought to have all the same beliefs, and in fact *every* pair of people ought to have the same credences! This is not in fact possible, given that a person is a member of many different groups, and these different groups already have different degrees of belief.

7.2.2 Shared interests

A more plausible version of this view proceeds from the features that later stages and earlier stages of an entity share in common. I have said that a doxastic state is rational to the extent that it does a good job of playing its functional role. This functional role could be alethic or pragmatic. If the earlier stages and the later stages of the entity are both part of what this doxastic state is functioning to govern, and if those alethic or pragmatic values can be promoted by behaviors or doxastic states that occur at multiple times, then there will clearly be reason for these stages to cohere.

In effect, to take this response is to do a modus ponens where Christensen wants us to do a modus tollens. He says that if the successive stages of a person have rational reason to cohere, then so do spouses that share a bank account. Since he rejects the latter, he rejects the former as well. However, I suggest that we embrace both, at least to the extent that there really are shared interests. The reason why we don't ordinarily think that it is a rational failing for spouses to have different degrees of belief is twofold.

First, there are just fewer lines of communication between a person and their spouse than between an earlier stage and a later stage of the same person. When lack of communication (including, in an individual human, forgetting) gets in the way of satisfaction of requirements of rationality, we tend to feel it is excusable, and a mere tragedy rather than irrationality. (This is what we often say about non-ideal subjects generally, as I will return to in Section 8.5.) Since even the closest spouses don't finish each other's sentences as well as the earlier and later stages of the same human body, there is more room for us to forgive failings of rationality for these groups.

Second, even the closest spouses don't share as many desires and values as the earlier and later stages of a typical human body, at least during the time frames that are typical for updates by conditionalization. While sharing a residence and sharing a bank account and sharing a family lines up a lot of the desires of spouses, and their values usually need to be quite closely aligned even before this for them to become spouses, it still just isn't as intimate and close a connection as the successive stages of a person. While traditional Christian conceptions of marriage say that the spouses "become one flesh", most modern understandings allow greater room for individuality in a marriage. But this distinction is not one of principle, merely one of degree—if multiple humans really do become as close in their values and their communications as the successive stages of a human body, then on this picture, they would be subject to the same sorts of requirements of rationality. And conversely, if a person undergoes what L.A. Paul (2014a) calls a "transformative experience", so that their desires and values change, then they will not be subject to a requirement of diachronic coherence for their degrees of belief across this change.

When Bayesianism is considered as a theory of entities other than a human, I think this sort of approach may be more useful. If a corporation, or an artificial intelligence, has very stable goals over time, and a means for tracking its behavioral dispositions and shaping them towards these goals, then Bayesian conditionalization is going to be an effective way to do so. But if the goals aren't stable, and means of communication aren't robust, then Bayesianism is likely to be a bad account of its diachronic behavior.

A similar version of this idea might be run on the alethic account of degree of belief proposed in Chapter 3. If the successive stages of a human being see their degrees of belief as collectively aiming at the truth in a single project, then the considerations of Greaves and Wallace (2006) or Briggs and Pettigrew (2016) will apply, and the stages will have reason to obey the rule of conditionalization. But the same can be true for collections of people, whether scientific research labs, or religious communities, or (conceivably) a family. This may be a very natural way to proceed for statisticians interested in motivating Bayesianism as a normative model for reasoning conducted by the scientific community as a whole.

This idea of the shared source of diachronic and interpersonal coherence is discussed further in Easwaran and Stern (2026). Similar ideas are discussed by Simon Evnine (2003, 2005). Some versions of this idea might well be paired with Parfit's contention that identity of the self across time is not all-or-nothing. However, instead of adopting attenuated prudential norms of concern for one's partial selves, or adopting impersonal norms of care for them as others, this picture *derives* the partial identity from the partially shared desires and values. It's not that one *should* partially desire what someone who is partially you desires, it's that what it *means* to partially be the same person is to desire many of the same things, and this give you reason to have coordinated credences that further strengthen the connection.

7.2.3 Korsgaard and the diachronic self

Christine Korsgaard (1989) suggests a different, Kantian, response to these considerations. Although the physical and mental substrate of our lives may not have true conditions of identity across time, but merely greater and lesser degrees of connectivity, Korsgaard notes that most actions are necessarily extended in time. For the Kantian, agency, in order to count as agency, is necessarily unitary, even if it is manifested in bodies and minds that are dissociated to greater or lesser degree. This is true whether the action and expression of agency is as brief as intentionally drinking a sip of water or as long as an intentional plan for how to live one's life. This can even be true for collective actions taken by a family, or a club, or a state, provided that they are acting intentionally as a unit (even as each individual may have her own separate actions done in parallel with the group action). Rather than any particular degree of physical or psychological continuity grounding a notion of personal identity sufficient to give us *reasons* for acting in a coherent way across time, she suggests that it is the unified action across time that *constitutes* the identity of the person. "You normally think you lead one continuing life because you are one person, but according to this argument the truth is the reverse" (p. 113). We might extend this sort of thought to states like degree of belief and desire, and not just action.

In the previous chapter, I suggested that the accuracy and pragmatic arguments for conditionalization could only justify the synchronic rules relating conditional to unconditional probability, together with a commitment to a plan for change in light of hypothetical future evidence. On a Korsgaardian view, we might say that the actually diachronic rule is just part of what it takes to *be* a unified agent across time. This could be read in a strong way or a weak way. On the strong reading, we might say that the epistemic self is extended through time only if the degrees of belief in fact change in line with the commitments of the earlier self. On the weak reading, we might allow that factors other than the way degrees of belief change over time are relevant to partial determination of identity as well. The strong reading would say that an agent who fails to update in line with conditionalization (or reflection, or whatever the correct version of the diachronic norm turns out to be) has not failed in any normatively significant way—there's just an epistemically different person present at the later time than the earlier time. The weak reading would allow that other factors could determine a degree of partial identity across time, which allows room for these factors to determine some normative reason for the actual update.

One important feature of this way of grounding the diachronic rule is that it explains how a rule for *change* of degree of belief can be important for *identity* across time. Christensen raised a worry that a Dutch book argument apparently similar to that for conditionalization can also be given for a rule to *never* change one's degrees of belief. (If one finds different prices for the same bet fair at different times, then buying the bet at the more expensive time and selling it at the cheaper time produces a guaranteed net loss.) However, when considered in terms of *commitment*, this argument is relevantly different from the ones for conditionalization and reflection. One shouldn't be committed to a *particular* change in one's degrees of belief, but instead should be committed to a *plan* that changes one's degrees of belief in one direction in light of one particular piece of evidence one might receive, and changes them differently in light of another piece of evidence one might receive. On this view, the criterion for epistemic identity across times involves making good on one's epistemic commitments, including commitments to change, more than direct psychological continuity.

Korsgaard doesn't consider this particular sort of change, but she does consider how agency can be unified across many psychological changes in the person.

> For the Kantian it does matter who is initiating the use of the mechanism. Where I change myself, the sort of continuity needed for identity may be preserved, even if I become very different. Where I am changed by wholly external forces, it is not. This is because the sort of continuity needed for what matters to me in my own personal identity essentially involves my agency. (p. 123)

If one's degrees of belief update in light of new evidence in line with one's earlier commitments, then one still has the unity of agency that the Bayesian picture describes. This can be true even when the evidence and commitment give rise to very drastic changes in one's degrees of belief (as is possible when one gains a piece of evidence that one antecedently considered extremely unlikely). Conversely, someone who gets this surprising evidence but somehow keeps her degrees of belief very similar to her pre-update degrees of belief has violated her commitment to conditionalize, and is thus in the relevant sense a different person after the update, despite some sort of similarity that she bears to the prior self.

7.2.4 Conclusion of diachronic norms

Michael Titelbaum (2012), raises an important question in Chapter 7 of *Quitting Certainties.* If, as I have suggested so far (and will argue more strongly in Chapter 8), the synchronic norms on degree of belief are just that they satisfy probabilism, and some amount of logical omniscience and a plan to conditionalize, then there are many different probability functions that an agent could have at a time, consistent with rationality. The diachronic rule says to change one's degrees of belief only if one gets new evidence. But if, without any new evidence, someone's degrees of belief happen to change from one coherent probability function to another, in what sense is this agent criticizable? (Recall that at the end of Chapter 4 I suggested that just this kind of vacillation between different coherent probability functions is the explanation of some cases that are sometimes described in terms of imprecise probabilities.)

The Parfitian view says that there is reason to cohere with everyone—not just one's past and future selves, but others as well. As long as there are others that the new self coheres with, it's not especially important whether the new self coheres with the old self.

The shared interests view says something a bit more specific. To the extent that the past and future selves have important shared interests in common, and to the extent that the pragmatic or epistemic behaviors relevant to those shared interests depend on credences in the propositions in question, a person who changes from one coherent state to another is irrational. But if the propositions about which one vacillates are not particularly significant for one's shared interests over time, then the degree of irrationality manifested here may be less than the many other types of irrationality than any actual person manifests through failures of logical omniscience and the like.

On the Korsgaardian view, we can say that if her plan was to maintain the same degrees of belief, but that her degrees of belief nevertheless change, then there is some sense in which she is just a somewhat different person afterwards than she was before. She is not irrational, but just a partially new person. She loses out on some of the values of extended epistemic agency.

On the other hand, if she had *planned* to change her degrees of belief like this, then she *is* one person through this change, but her plan was irrational in whatever sense a violation of the synchronic norms is irrational. (This will be investigated in the next section.) Oddly enough, if she plans to change her degrees of belief to another coherent probability function, but she actually maintains her current degrees of belief, then this view says that she is now a different person, despite nothing changing! The new self is rational (provided that she doesn't have any strange plans of this sort), but the old self was irrational (because of the plan).

There are two types of failure—planning and execution. These can be seen as failure of commitment to unity on each end. On my interpretation, the planning failures are synchronic failures of rationality of the same sort as violations of the probability axioms. (We will investigate these in the next section.) The failures of execution however, on this Korsgaard-type picture, just constitute non-identity over time. One isn't irrational—one just isn't truly the same person in some relevant sense. This may make things work out badly for the successive occupants of the body in some ways (or it may make things work out well).

7.3 RETURN TO SYNCHRONIC NORMS

In light of the split brain cases and others, we might want to say something similar about synchronic rules as diachronic rules. Perhaps the axioms of probability theory and the synchronic rules about how to plan to update aren't *normative* on agents, but instead are *constitutive* of being one agent in the first place. The people described in the previous section, with plans to update that don't cohere with their current degrees of belief, aren't irrational, but are rather better thought of as disunified person-parts that happen to share a body at a time.

Something similar is proposed by David Lewis (1982) for people with inconsistent beliefs.

> I speak from experience as the repository of a mildly inconsistent corpus. I used to think that Nassau Street ran roughly east-west; that the railroad nearby ran roughly north-south; and that the two were roughly parallel. (By "roughly" I mean "to within 20°".) ...What about the blatantly inconsistent conjunction of the three sentences? I say that it was not true according to my beliefs. My system of beliefs was broken into (overlapping) fragments. Different fragments came into action in different situations, and the whole system of beliefs never manifested itself all at once. The first and second sentences in the inconsistent triple belonged to—were true according to—different fragments; the third belonged to both. The inconsistent conjunction of all three did not belong to, was in in no way implied by, and was not true according to, any one fragment. (p. 436)

Similarly, a person with an incoherent set of degrees of belief may well be better understood as having several fragments, each of which is coherent. This view is developed further by Elga and Rayo (2021a,b).

Furthermore, the person with one coherent probability function giving her current degrees of belief, and the plan to update her degrees of belief to a different coherent probability function without getting any new evidence, is perhaps better thought of as two fragments of a person. If she does in fact update to the other probability function, then the second fragment is the one that survives to the future person. If she in fact stays with the original probability function, then we might say that the first fragment survives, though the lack of a plan for update in that first fragment might mean that the later person is not the same person as either earlier fragment, but is just similar to the first fragment. On the Korsgaardian view, having plans is partially constitutive of diachronic identity, so a fragment must contain plans in order to survive through time at all.

The arguments for the probability axioms might be taken in a way that supports this view. Consider Ramsey's justification of the Dutch book argument (as in Chapter 2) for the claim that degrees of belief must satisfy the probability axioms:

> Any definite set of degrees of belief which broke them would be inconsistent in the sense that it violated the laws of preference between options, such as that preferability is a transitive asymmetrical relation, and that if α is preferable to β, β for certain cannot be preferable to α if p, β if not-p. If anyone's mental condition violated these laws, his choice would depend on the precise form in which the options were offered him, which would be absurd. He could have a book made against him by a cunning better and would then stand to lose in any event. (Ramsey, 1926, p. 182)

We can read Ramsey as saying that what it is for certain metal states to count as preferences is in part for those states to satisfy transitivity, and for those states not to favor an act with a possibility of a worse outcome and no possibility of a better one over an act that is guaranteed to have a better outcome. Note that he doesn't say that it is *irrational* or *unreasonable* for someone's choices between options to depend on the form in which they are presented, but that it is *absurd*.

This interpretation is only plausible if the notion of possibility at work here is one that is in some sense part of the mental states themselves. One has maximal degree of belief in something that is true in every possibility, and if A and B are disjoint sets of possibilities, then one's degree of belief in $A \cup B$ just is the sum of one's degrees of belief in A and in B. This is plausible if the notion of "possibility" is the agent's own notion, but not if it is meant to be a kind of logical possibility, as is often assumed in discussions of Bayesianism. This is why, throughout this book, I have emphasized *doxastic* possibility, though it might be more relevant to follow Hacking (1967b) and call it "personal possibility".

This doesn't mean that incoherent degrees of belief are impossible within what appears to the world to be one person. Because of the possibility of fragmentation, as mentioned by Lewis, we could well find mental states residing within one head that fail to satisfy the probability axioms. However, this would suggest that each fragment would have its own relatively well-developed set of propositions to which it had degrees of belief. It doesn't seem plausible that one fragment could have a degree of belief in A but no degree of belief in $\neg A$, while the other fragment has a degree of belief in $\neg A$ and no degree of belief in A. Each fragment could have its own domain of propositions over which it has degrees of belief, but these will likely need to consist of many propositions as well as ones that are closely conceptually connected to them, rather than just a few isolated propositions here and there.

Lewis goes on to say, "Once the fragmentation was healed, straightway my beliefs changed: now I think that Nassau Street and the railroad both run roughly northeast-southwest." It's not clear to me what it means to "heal" a fragmentation, or how we should think of the revision involved in doing so. There may be norms here. If so, they may connect to the literature on disagreement—when ideas come into contact, there is a reason for them to cohere. We feel less pressure across individuals or across times than we do within an individual, but this may be a prejudice that is supported by the lack of constant two-way contact.

7.4 CONCLUSION

There are three important types of norms that are often taken to be constitutive of Bayesianism. There is the synchronic norm of logical omniscience, the quasi-diachronic and diachronic norms of conditionalization, and the synchronic norms of probabilism. I argued that these have three distinct sources of justification. As long as we allow the doxastic possibilities for the subject to include situations that are actually impossible, the standard arguments do not justify logical omniscience. Instead, logical omniscience must follow from a requirement to manage these doxastic possibilities appropriately. The diachronic norms depend on the nature of the unity of the subject over time and their commitment to collective coherence. Parfit suggests that there is in fact some sort of pressure of unity for *all* subjects, while Korsgaard says there is a distinctive feature of unity that is essential to Kantian agency. I suggest that it derives from the pragmatic or alethic commitment to shared values, which might apply across groups just as well as across times. But probabilism itself just follows from the concept of pragmatic or alethic value, and the subject's own concept of possibility. To the extent that a subject fails to live up to this, we might re-interpret the subject as several fragments with only a loose commitment to each other.

Research Questions

1. What is the status of norms of logical omniscience?
2. Is the impossible worlds framework an appropriate way to account for logical non-omniscience?
3. How can we compare the irrationality of violating probabilism, logical omniscience, or conditionalization? Or are these incomparable in some way?
4. Is it best to understand diachronic norms by thinking of a diachronic agent as a group of synchronic agents, or in terms of some sort of Kantian commitment, or in some other way?

CHAPTER 8

Multiple rationalities

IN THIS CHAPTER, I consider the concept of rationality for degrees of belief more generally. I take a pragmatist stance. A cognitive state is rational to the extent that it does a good job of playing its functional role. An account of rationality for degree of belief must follow from the central role degree of belief plays (whether pragmatic or alethic). But I will argue that there are many different "perspectives" one can take when evaluating what it takes to do a "good job" of playing this role. I argue that different perspectives yield different overall scales of rationality. Many disputes about epistemic rationality are thus, I claim, terminological disputes, about which perspective is the "right" one.

From one very natural perspective, perfect rationality will consist in satisfying probabilism, conditionalization, and logical omniscience, yielding a very permissive conception of rationality. But from other perspectives, there could well be additional norms, and a doxastic state that falls short of some of these three basic norms might actually do better overall than one that satisfies them. These contrasting views of rationality are connected to the distinction between an "internal" and an "ecological" conception of rationality, the question of whether individual attitudes or policies for setting attitudes are the right level of evaluation, and the contrast between ideal and non-ideal theories of rationality. I argue that epistemology, even for a purely solitary agent, needs to consider issues from social and political theory.

8.1 PERMISSIVE RATIONALITY

From the internal perspective of an agent, violation of probabilism is irrational. All of the arguments in Chapter 2 use features of the agent's

DOI: 10.4324/9781315626918-8

preferences to construct numerical representations of their degrees of belief and utilities. If the agent satisfies enough of the axioms to ground these constructions, then if these degrees of belief failed to satisfy probabilism, there would be cases where the agent prefers one thing to another, even though they see no possibility of the former being better than the latter, on their own utility scale. If credences are conceived of in the forms from Chapter 3, as essentially aiming at the truth, then having credences that violate probabilism means that there is another set of credences for which the subject sees no possibility of their actual credences being as accurate as these alternates, so by their own lights, the subject is not as accurate as they could be. The various structural axioms that set up the different representation theorems, and the different sets of constraints on measures of accuracy, are best thought of as constitutive of the relevant conception of doxastic subject in question—a creature that doesn't satisfy these actions might be a different type of doxastic subject, or might not be a doxastic subject at all. But given that they are a doxastic subject of the relevant sort, the violation of probabilism shows that their doxastic state is defective by its own lights.

8.1.1 Pragmatic and alethic routes to probabilism, but nothing more

I claim that if a doxastic subject's doxastic state, as numerically represented in the relevant way, *does* satisfy probabilism, there is no such internal criticism available. For any two different probabilistic doxastic states over the same set of possibilities, there are always possibilities where one does better (by either pragmatic or alethic evaluation) and possibilities where the other does better (by the same sort of evaluation).

On the pragmatic conception of the doxastic state, for any two distinct probabilistic doxastic states, there is a pair of acts such that one doxastic state prefers the first act, and the other doxastic state prefers the second. On the Dutch book argument, and Ramsey and Savage's decision theories, everyone prefers dominant acts, so neither of these acts can dominate—there must be some possibilities both agents recognize where one act does better, and some possibilities both agents recognize where the other act does better. On evidential decision theory, the agent's act partly determines which world they end up in, so there isn't some *world* where one set of degrees of belief would lead to an act with a better outcome, and some other *world* where the other degrees of belief would lead to an act with a better outcome. But nevertheless, if one has higher evidential expected utility for one doxastic state, and another act does for the other, even though the utilities of the consequences are the same, then both doxastic state must recognize the possibility of one act turning out well, and the possibility of the other act turning out well. There is no agreed-upon criticism of one act until after the results are known—and regardless of how it comes out, the agent whose preferred act turned out worse can still point out that the other wasn't *certain* it would in advance. Similarly for causal or functional decision theories—even for agents who share the same utilities, the differences between different doxastic states that satisfy the probability axioms always *could* lead to one doing pragmatically better or the other. Thus, on a pragmatic conception of degrees of belief, there can be no definitive criticism of one probability function over another, though there are definitive criticisms of non-probabilistic doxastic states. Each one might dismiss the other as doing worse in more important worlds and better in less important worlds—but the way that each measures importance is by its own probabilities, which are precisely what is at issue.

Similarly, on the alethic conception of the doxastic state, for any two distinct probabilistic doxastic states, there is a pair of worlds such that one doxastic state is more inaccurate in one world, and one doxastic state is more inaccurate in the other. This is because the scoring rule is proper (as justified in Section 4.1.3.3). If one of the doxastic states were more inaccurate than the other in every world, it would have higher expected inaccuracy even calculated using its own probabilities. But since the scoring rule is proper, each of them must have some world where it is less inaccurate than the other. Again, each might say that the other is only doing better in less important possibilities, but the importance of these possibilities is precisely what is at issue in the debate between different probability functions.

8.1.2 A more substantive conception of rationality?

Many philosophers think that there should be a more substantive conception of rationality. As an example:

> Reducing rationality of utility to internal consistency may result from the ancient recognition that tastes are inherently subjective (de gustibus non est disputandum). But one may say more about beliefs than about tastes. Rationality does not constrain one to like or to dislike the smell of tobacco but rationality does preclude the belief that smoking has no negative health effects. (Gilboa et al., 2012, p. 13)

Some go so far as to say that given one's evidence, rationality in fact puts a *unique* constraint on what credence function someone may have. Horowitz and Dogramaci (2016) argue that rationality is inherently a communal concept that is used to encourage others to reason in compatible ways, and that it can only serve this purpose if there is a single way of converting evidence into degrees of belief that counts as rational. But even opponents of this sort of uniqueness, such as Schoenfield (2014), suggest that there are rational requirements on credence beyond mere probabilism.

I will allow that there is in fact a conception of rationality that puts stronger requirements on credence, but argue that such a conception of rationality is ill-suited for abstract philosophical theorizing. Such a conception of rationality will depend extensively on empirical facts that philosophers are not in a good position to reason about. It will also put further requirements of rationality on *taste* and not just on belief. The conception of rationality that is most amenable to abstract philosophical work includes no requirements beyond probabilism. The distinction between these two conceptions of rationality seems to have been fairly clearly understood by the early pioneers of modern Bayesianism. (See Sections 4 and 5 of Ramsey (1926) and all of de Finetti (1931/1989).)

Consider what goes on in a case where one person is highly confident in the proposition that smoking causes lung cancer, while another is not. Let a be the proposition that smoking causes lung cancer, and let P_1 be the first person's degrees of confidence (the ones we consider reasonable), while P_2 is the second (the ones Gilboa, Postlewaite, and Schmeidler say is irrational). The naive thought is that the first person is more rational than the second, because the second is more likely to get cancer. However, as I will discuss in greater depth later, any response like this will end up making a parallel judgment about tastes. A person who doesn't like the taste of cigarettes is much less likely to end up getting cancer than someone who does like the taste of cigarettes. But I will argue that the only way to argue that there is something irrational about being confident that smoking does not cause lung cancer is going to end up committed to this same sort of irrationality of tastes.

8.1.3 Considerations of evidence

First, note that, considered in pure abstraction, most philosophers will say there is nothing inherently wrong with $P_1(a)$ being high while $P_2(a)$ is low—the problem only arises when we specify that both people have access to the same evidence that we do. The main relevant evidence that we have consists of decades of reports of lung cancer. Let b be the proposition that summarizes all the relevant data someone has. It is quite clear that everyone in the contemporary world has somewhat different data. For many of us, b consists primarily in the fact that we have heard that smokers get lung cancer much more often than non-smokers. For others that have known particular friends and family that have gotten lung cancer and/or been smokers, we have information about those specific cases. Many doctors have much more detailed data about hundreds or thousands of cases of lung cancer and smoking habits of some of those people.

One natural way to think about the effect of this evidence is to imagine the hypothetical degrees of belief P'_1 or P'_2 that each person would have had *without* this evidence, and to consider the way these degrees of belief would have changed in response to b. That is, we should have $P_1(a) = P'_1(a|b)$ and $P_2(a) = P'_2(a|b)$. As mentioned in Chapter 7, it is often easier to consider the force of evidence on the odds representation rather than the probability representation. Let $O'(a) = P'(a)/P'(\neg a)$ and $O(a) = P(a)/P(\neg a) = P(a|b)/P(\neg a|b)$. As mentioned in Chapter 7, this means (assuming the person obeys conditionalization, and logical omniscience for very simple conjunctions) that

$$O(a) = O'(a) \cdot \frac{P'(b|a)}{P'(b|\neg a)}.$$

To try to argue that one person is irrational here, we can look at what the people share in common for this formula, to see how the other part would have to differ in order to rationalize their different current beliefs about smoking and cancer. Although b might constitute a different proposition for each of us, there are some features that it seems to have in common for all of us. For the lucky people that know about lung cancer only in the abstract, our evidence b is the proposition that smokers get lung cancer much more often than non-smokers. This is not the same thing as a, but it is quite plausible that $P'(b|a)$ is very high, while $P'(b|\neg a)$ is very low. For the people who have more detailed personal acquaintance with lung cancer cases, b will contain many specific claims about specific individuals getting lung cancer and other individuals not, and any such particular proposition b will have $P'(b|a)$ being fairly low. (Even if smoking causes lung cancer, it is highly unlikely that these *particular* three smokers will get cancer on the particular dates they do, while two other smokers never do, or whatever the particular pattern of observations is.) However, $P'(b|\neg a)$ will be even lower. For anyone with a reasonable understanding of what it is to "cause cancer", and any conception of the relevant evidence b (assuming that it involves far more cancer for smokers than for non-smokers), it seems quite plausible that $P'(b|a)/P'(b|\neg a)$ will be very high.

If the person's prior odds had been less extreme than this value of the likelihood ratio, $P'(b|a)/P'(b|\neg a)$, then we can say that learning via conditionalization would have led them to become confident that smoking causes lung cancer, after observing this evidence. So if they aren't confident, despite having the evidence, then they must have violated either the integrity requirements that lead to conditionalization, or the very simple logical requirements on a conjunction. Thus, by their own lights, there would in fact be something irrational about having this low credence in smoking causing lung cancer.

However, no matter how high the likelihood ratio is, there will also be *some* prior degrees of belief P' that leave the posterior odds quite low! If we imagine a likelihood ratio of a million to one (that is, the evidence we've observed is a million times more likely given that smoking causes lung cancer than otherwise), then anyone whose prior credence in smoking causing cancer was above 0.000001 would end up confident that smoking does cause lung cancer. But nothing has yet been said to argue that having such a low prior credence in a hypothesis absent relevant evidence is irrational. Lung cancer is a very specific range of outcomes, and it's not clear that anyone *should* find it *at all* plausible that any particular behavior would cause this range of outcomes!

8.1.4 A biased coin vs. a biased urn

As an analogy, consider the difference between two formally very similar cases. In one case, we have an urn that is labeled as containing only black and white balls, but with no indication of how many of each. In the other case, we have a coin that clearly has a heads side and a tails side, but with no indication of how it is biased. Before we begin, consider your prior that an actual coin is biased, and compare it to your prior that an actual urn has a different number of black and white balls in it.

Imagine that we make 5 flips of the coin (with an assistant catching the coin in mid-air), and it happens to come up heads each time. This might lead us to a *slight* suspicion that the coin is biased. Imagine further that we make 5 draws from the urn (replacing the ball and shaking up the urn each time), and we happen to get a black ball each time. This would likely lead us to a much stronger suspicion that the balls in the urn are mostly black.

After 20 such draws or flips, we would be almost certain that the urn is mostly black, but in the case of the coin, I suspect that most of us would still think there is a serious possibility that this is just a fluke of luck.

This is largely because it's quite easy to imagine how to bias the urn—just adding more black balls and fewer white balls would bias it. Meanwhile, it is quite hard to imagine how to bias a coin, and I confidently predict that you yourself have never actually encountered a biased coin. (In fact, Gelman and Nolan (2002) give simple physical arguments and classroom experiments intended to show that it is *physically impossible* to bias a coin that is flipped and caught in midair, even though it is quite easy to make a loaded die.)

The likelihood ratio of 20 consecutive heads, or black draws, is a million to one in favor of the extreme biased hypothesis over the fair hypothesis. We can easily feel how the priors of a biased coin and the priors of a biased urn are different enough to make these posteriors quite different. I claim that there is no good reason why priors about the causal structure of smoking and cancer, or vaccines and autism, or creation and evolution, shouldn't be even more different.

8.1.5 Irrational, or merely extremely unusual?

However, in some of these cases, there is actually a lot of public evidence that is in fact shared. (This may be the case somewhat less often than we expect, because we all live in our own "epistemic echo chambers" (Nguyen, 2020), but for some of these disputes, the evidence is in fact shared.) In order to sustain the different posteriors, people would have to have *extremely* different priors, with not just a million to one difference in odds, but billions or trillions.

My claim is that in these cases, such extreme priors are extremely *unusual*, (just as it is unusual to find it more likely that an unknown coin is biased than that an unknown urn has a different number of black and white balls in it) but they are not actually *irrational*. Some philosophers have noted that we *want* to judge people as irrational on the basis of having sufficiently different posteriors in these cases, and have used this as an argument against the kind of radical permissivism I endorse. After all, my permissivism says that these people *could* be rational, provided their priors were extreme enough.

However, I think this argument fails. As evaluators of others, we should be Bayesian ourselves. I acknowledge that there *could* be perfectly rational people with very extreme priors, who end up with these posteriors after evaluating the evidence. But I also acknowledge that there *are* many *irrational* people who don't update their beliefs in line with conditionalization. When I encounter someone whose posterior is unusual in this way, I consider both of these possibilities, and my priors for each.

Although there are *some* priors that would lead to a rationally low posterior credence in smoking causing lung cancer, such priors are rare, and I may never have encountered someone with such priors. However, I *have* met many people who update irrationally—at one point in time, they have degrees of belief that lead them to say that $P'(b|a)/P'(b|\neg a)$ is much more extreme than their current odds $P'(a)/P'(\neg a)$, but due to motivated reasoning of one sort or another, after learning b they stubbornly continue to think a is unlikely.

William Miller's interpretations of scripture predicted, with great confidence, that Jesus Christ would return to the world on October 22, 1844. In the following months, there were a variety of reactions among Miller's followers. Some used the "Great Disappointment" to give up on their beliefs in Miller's interpretation. But others had other reactions. Some reinterpreted the date not as the date of the second coming, but rather the date of a heavenly "door being shut", after which conversion would no longer be effective for salvation. But over time, the fact that there was no observable event on that date led many to adopt one of two other interpretations—either the date had been calculated wrong, or in fact Jesus *had* returned, but invisibly. Some individuals may have had prior probabilities that lined up with the interpretation they eventually gave of the Great Disappointment. But the fact that many of them switched groups, and that the last two interpretations ended up being most common, leads me to believe that most of these people were revising their beliefs under social pressure and the influence of "cognitive dissonance", rather than in line with the priors that they had been committed to.

Because this sort of irrational updating seems to be much more common than the combination of rational updating with a very extreme prior, I find it likely that the person in the cancer case is being irrational. This is particularly true when a is something emotionally or politically salient for which it is nevertheless easy to gather even more evidence pointing towards the scientific consensus—smoking causing cancer, the development of life through Darwinian evolution, the effectiveness and safety of vaccination. I claim that most people who have the minority judgments in these cases really are being irrational by their own standards in not updating via conditionalization. (This can be seen by the fact that their credences in the scientific consensus appear to remain moderately low, around 0.001, regardless of how much more evidence comes in, rather than moving between 0.00000001 and 0.00001 or the like.)

I claim that it is abstractly *possible* for someone to have prior judgments that *would* have counted the posteriors as rational. But actual people almost never exhibit such extreme priors. So it is reasonable to judge anyone with these posteriors as irrational. There is no need for a radical principle of charity that always evaluates everyone as rational if there is any possibility of a rational prior leading to those posteriors. Rather, I am committed to there being some number of people with priors somewhat dissimilar from my own, who update irrationally and a much smaller number of people who are perfectly rational and have priors extremely dissimilar from my own. So if someone ends up with a posterior that is sufficiently radically different from my own, despite apparently having the same evidence, it is much more likely that they have updated irrationally than that they have updated rationally from an unusual prior.

Of course, *this* argument *itself* will only convince you if you share my priors about the relative probability of a person having an unusual (but, I claim, rationally permissible) prior, as compared to a person updating irrationally from a more common prior.

(One other implication of my views here—most people who share my political, scientific, religious, and other beliefs are *also* irrational. Irrational updating is quite common. I just have less glaring evidence of it. And to the extent that my worldview is not based on claims that are easily proven false, my own irrational dispositions *may* be less likely to be activated.)

8.2 ECOLOGICAL RATIONALITY OF POLICIES

My justification for the claim that nothing beyond probabilism was rationally required depended on an internal assessment of the subject's doxastic state. If someone violates probabilism, then there is some alternative doxastic state they could have had, that they would recognize as doing better by their own values (whether pragmatic or alethic) than their own doxastic state, in every possibility they recognize. But for any proposed principle beyond probabilism, the subject can point to some possibility on which their own state does better by their lights than ones that satisfy this proposed additional rule. They thus have an internal response to any claimed rules beyond probabilism.

However, when evaluating third parties, we don't always hold our evaluations to their internal standards. After all, the *point* of rationality is not so that the subject can be happy with their own doxastic state—the point is to *actually* do well (either engage in successful action, or be accurate in one's representation of the world). Internal evaluation is just the most that the subject can give from their own position. If they can internally tell that their doxastic state is worse than another, then there is an alternative that they can recognize as an improvement—but there will of course be many changes that are actually improvements beyond the ones that the subject can already recognize to be improvements.

Things are likely to go badly for a person who has low credence that smoking causes cancer. From an alethic perspective, this credence is itself highly inaccurate. And from a pragmatic perspective, it might be worse (at least, assuming the person has normal preferences about how badly they don't want cancer).

However, this is just a single low credence in a proposition that is in fact true. This is not usually what we would consider irrational. In the case where the person making the judgments is a doctor, for whom b is a highly specific pattern of particular details of cancer cases over many different patients, it is quite reasonable to assume that any ordinary doctor would have had extremely low credence in b before observing the actual patients. (To avoid having such a low prior credence in the actual distribution of particular facts about their patients, the doctor would have had to exhibit preternatural foresight about precisely which symptom each patient would develop at which time!) And yet we don't want to say that an ordinary doctor is irrational for having such low credence in the truth.

On the pragmatic account, it's easy to draw a distinction between the degree of belief about smoking causing cancer, and the degree of belief about the particular pattern of cancer cases—the former, but not the latter, is directly relevant to some highly practically significant decisions. But it would be better to have a response as to why a single low degree of belief in a proposition that is in fact true is not by itself sufficient for irrationality.

8.2.1 "Nearby" possibilities

A natural thing to note here is that although the doctor's degree of belief in this particular pattern of cancer cases is low, and thus is in fact inaccurate when this is in fact the pattern of cases, it is not *robustly* inaccurate. There are a lot of nearby possibilities where this low credence is in fact highly accurate, and there are also a lot of nearby propositions whose low credence is in fact highly accurate. But with the low degree of belief in smoking causing cancer, there seem not to be many (if any) nearby possibilities where the low credence is accurate, and there seem not to be many (if any) nearby propositions whose low credence is accurate. Furthermore, even if the nearby possibilities have different precise patterns of cases among patients, they would still almost all have been taken by anyone with normal priors to provide similar amounts of evidence in favor of the cancer hypothesis. (I will not investigate precisely what "nearby" might mean here—different philosophers have used the concept in different ways.)

Many epistemologists have claimed that these sorts of nearby possibilities have some normative significance for full belief and its relation to knowledge. Nozick (1981) argued that for a belief to count as knowledge, it must be "sensitive" (that is, one wouldn't have held the belief in the nearest possibilities where it was false), and "adherent" (that is, one would have held it in all the nearby possibilities where it was true). Sosa (1999) argued that instead it should be "safe" (that is, there are no nearby possibilities where one believes it and it is false). For more details on these sorts of conditions, see section 5 of Ichikawa and Steup (2017).

One might say something similar about degree of belief. On a pragmatic conception of the role of degree of belief, one might say that a set of degrees of belief is "robustly irrational" if they lead one to make a set of decisions on which there is no nearby possibility for one to come out ahead. On an alethic conception, one might say that a set of degrees of belief is "robustly irrational" if there is an alternative set of degrees of belief such that there is no nearby possibility in which the set one has is more accurate than the alternative. These two definitions replace the agent's own doxastic possibilities with the set of nearby possibilities. Thus, if we re-formulate the arguments from Chapters 2 and 3 in these terms, we can see what the result of such arguments must be. If X is the set of possibilities that go into any of these arguments, then the rational requirement that comes out is that the doxastic state should be a probability function over X. In the previous chapters, X was the subject's own doxastic possibilities, but now X is the set of "nearby" possibilities. Among the conclusions that follow, it turns out that one is irrational if one's own doxastic possibilities are not all and only the nearby possibilities. This implies logical and metaphysical omniscience as a byproduct, if "nearby possibilities" are all logically or metaphysically possible.

We might also consider probabilistic versions of these arguments, to weaken the omniscience requirement. Sherri Roush (2005) gives probabilistic versions of sensitivity and adherence for full belief, saying that for a belief to count as knowledge, it should be highly improbable that one believe it and it is false, or that it be true while one doesn't believe it. If there is some objective probability function over the nearby possibilities, then arguments like those in Pettigrew (2016a) may lead to the requirement that a rational agent have degrees of belief that match this objective probability function. In fact, a version of this view is suggested by Williamson (2002) and others.

8.2.2 Ecologically valid possibilities

However, I claim it is more useful to consider these ideas as part of an explanation of the *unusualness* of credence functions that these arguments purported to show to be irrational. Consider Figure 8.1. I have drawn it so that one of the lines is 10% longer than the other, as you can verify with a measuring implement. Most people will not be very confident in their judgment of which is longer.

An even more extreme version of this occurs in the "Ames room" (see Figure 8.2). Since our eyes are so used to seeing rectangular rooms

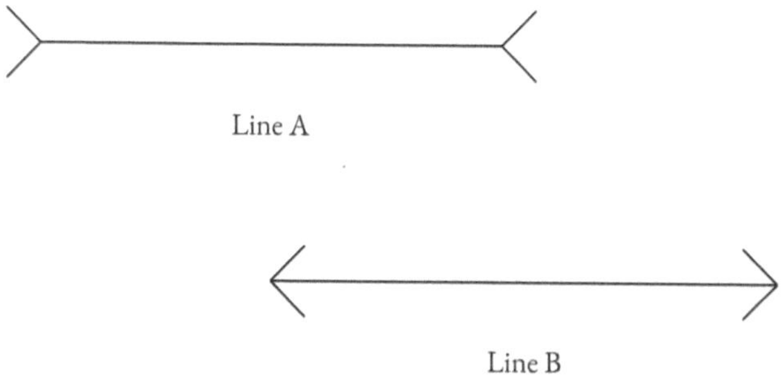

Figure 8.1 The Müller-Lyer Illusion. Lines A and B are *not* precisely the same length. One is 10% longer than the other.

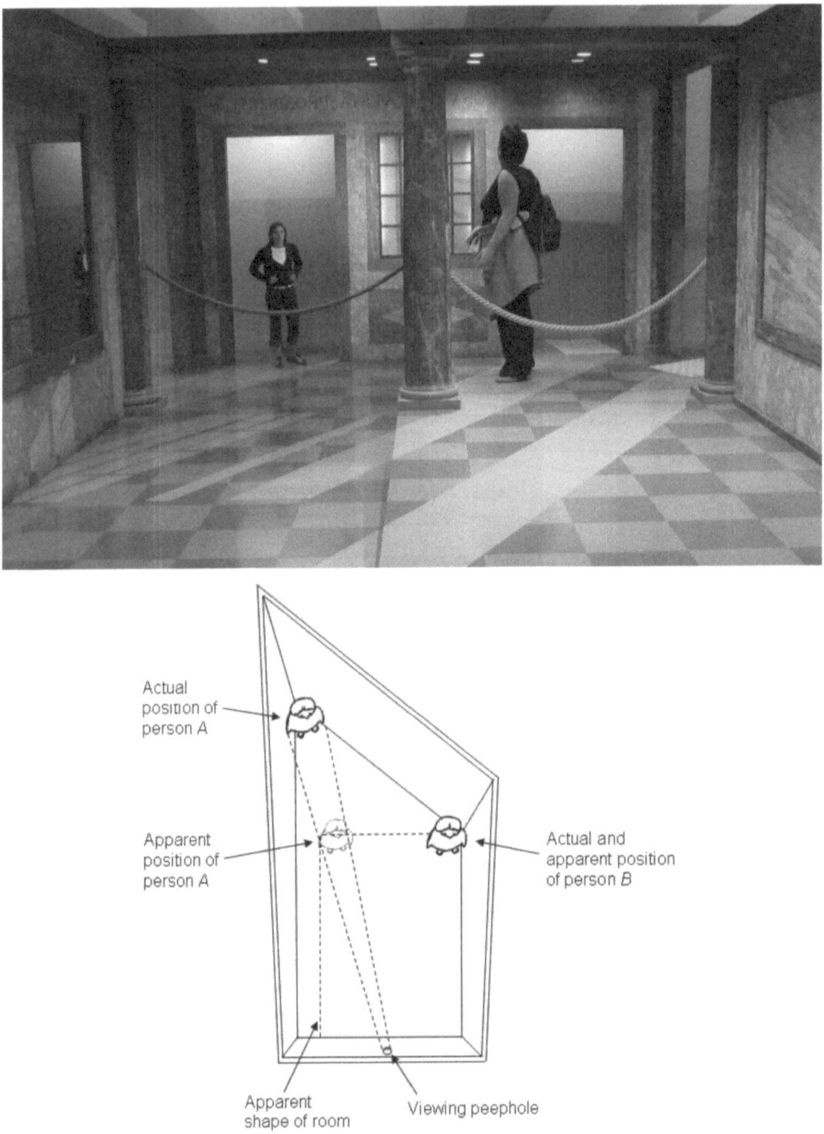

Figure 8.2 The Ames Room illusion. Photo CC by Mosso, diagram CC by Eixo, both from Wikipedia.

with square floor tiles, we interpret the image as such a room, and view the one person as much larger than the other, even though this ratio of heights is extremely unnatural. Technically, whenever our retinas get visual impressions as of a person in a rectangular room, those impressions are ambiguous between a room that is actually rectangular with an ordinary person in it, and a room that has this funny shape, with a person of unusual size. However, since these unusual rooms are so uncommon, our visual systems tend to produce more accurate judgments that are better guides to preference and action when they deliver a confident judgment of a rectangular room than when they deliver a judgment that each of these possibilities is equally likely. In the rare case that we actually do encounter the Ames room, such a visual system will lead us astray. Regardless of which possibilities are "nearby", only one set of possibilities is common in our home environment.

The explanation of the Müller-Lyer illusion is similar. In the environments that we tend to encounter in daily life, a line that is framed with inward-hooked lines is usually the front edge of some convex rectangular solid, while a line that is framed with outward-hooked lines is usually the back edge of some concave rectangular solid. Thus, lines framed with outward-hooked lines are usually farther than lines framed with inward-hooked lines, so if they occupy the same angle on the retinas, the former usually represent a longer line than the latter. Our visual systems do a very good job at judging the length of lines in real-world contexts, in the types of environments we tend to occupy, and a worse job at artificial tasks in the context of psychological illusions. (Howe and Purves, 2004) Thus, we might say that these features of the visual system that make us susceptible to these illusions are "ecologically functional"—they do well in our natural ecology.

I will follow Gigerenzer and Todd (1999) and Todd and Brighton (2015) in considering a parallel notion of "ecological rationality". ("Eco" comes from the Greek word "*οικος*", meaning "home".) On a pragmatic account, the role of an agent's doxastic state is to guide their actions. One doxastic state ecologically dominates another if it leads that agent, *in their home environment*, to perform actions that yield better consequences than the actions that the other credence function would lead them to perform. On an alethic account, the role of a subject's doxastic state is to aim at accuracy. One doxastic state is more ecologically rational for a subject than another if it has a higher accuracy over all possibilities *in the subject's home environment* than the other.

An abstract, un-located concept of rationality only requires probabilism, because a subject *could* find itself in any environment. For any two probability functions, there will be some environments in which one does better and other environments in which the other does. So ruling out some probability functions would make a person do worse in some environments. Any proposed definition of rationality that requires extra principles might lead the subject astray, if they find themself in one of these environments. But if the goal is only to do well in your *home* environment, rather than in *whatever* environment one might conceivably end up in, then something more specific might be reasonable.

And in fact, in many cases, if we are interested in success in particular environments, certain specific *violations* of Bayesian norms might do better than following Bayesian update procedures from reasonable-seeming priors. Douven (2013) has shown that, in a particular game of estimating the biases of coins, adding a factor to the update process to "speed up" conditionalization tends to help a subject get closer to the truth faster. On *average*, this rule did worse than the Bayesian rule, but in *most* cases it did a little better (and in a few cases did a lot worse). Depending on what goals are important for the subject, this may well be better performance.

One thing to note about these "ecological" notions is that there isn't obviously a single objective ecology. It turns out that the magnitude of the Müller-Lyer illusion varies across cultural groups. When the effect was tested in about 20 different populations, the effect was found to be strongest in people from western urban environments, and weaker in people from African urban environments, and weakest in rural African groups. (Segall et al., 1963) A different illusion involving faulty perception of length of a vertical and horizontal line showed the reverse pattern. It would be natural to suppose that the strength of these illusions is related to the frequency with which these various geometric shapes appear in the different environments different people live in—people surrounded by skyscrapers and tables and streets and cardboard boxes encounter a very different set of perpendicular lines than people surrounded by trees and traditional handmade dwellings. It has also been found that the illusion appears in artificial neural nets trained with various images, and certain higher-level structures in the neural net were found to control the magnitude of the effect (García-Garibay and de Lafuente, 2015).

This suggests that for agents that live in different environments, there is a differently relevant conception of ecological rationality. And even for a single agent, there are likely very different ranges that could be considered the "home"—we could be focusing on the built environment the agent lives in, their social environment, the ancestral evolutionary environment, or others. An agent that moves from one environment to another, or whose environment changes around them, likely undergoes a transition from one suite of ecological adaptations to another, though there is likely no specific time at which this transition is rationally required.

8.3 PHILOSOPHICAL CASE STUDIES

Thinking about cases from the philosophical literature through the lens of ecological rationality can help us understand what ecological rationality requires, and what is going on in these cases. I won't give a full evaluation of these cases, because these cases involve belief rather than confidence or degree of belief, and I don't have a functional role characterization for belief. But thinking about what it means for beliefs to function well in an environment helps clarify the contrasting views in these cases.

8.3.1 Fake barns

First, let us consider the "fake barn" case from Goldman (1976). In this case, Henry is driving through the countryside and sees what looks like a barn. Unbeknownst to him, he has entered a region where many farmers have erected barn facades that look from the road like barns. However, the one he is in fact looking at currently is actually a barn, even though there are many mere facades nearby.

Goldman's judgment is that Henry's belief that he is looking at a barn is justified, and is true, but is not knowledge. It is justified because it is generated in the same way that Henry's ordinary judgments about barns in ordinary locations is generated. It is true because he is in fact looking at a barn. But it is not knowledge, because Henry couldn't tell a facade from a real barn from the way he was looking, and in his present circumstances, the possibility of being a facade is a significant one that would have to be ruled out in order to have knowledge.

On a paradigm of ecological rationality, we can say something a bit more specific about Henry's belief. The method he uses to generate this belief is likely to lead to many true beliefs in his home environment. However, it is likely to lead to many false beliefs in his new environment. So considered from the perspective of the one environment, this method is ecologically rational, but considered from the perspective of the other, it is not. A method that proceeded from his sensory data to the judgment, "either there is a barn in front of me, or at least a facade of a barn", would yield true beliefs in both environments, but it would be less useful in his home environment.

These two methods can be further illustrated by means of a classic joke.

> A mathematician, a physicist, and an engineer were taking a train up to Scotland for a conference. Soon after the train crossed into Scotland, they looked out the window and saw a black sheep. The engineer said, "Aha! In Scotland, the sheep are black!" The physicist said, "Wait, wait—all we can say is that in Scotland, there is at least *one* black sheep." The mathematician said, "Well, not quite—in Scotland, there is at least one sheep, at least one *side* of which is black."

We can see that the engineer's dispositions are quite useful if one often travels from country to country, visiting places with monochromatic flocks of sheep. The engineer quickly forms a strong judgment and can use this to act confidently, though she is at risk of making a mistake in a place where sheep of multiple colors flock together. The physicist and the mathematician, however, make judgments that illustrate what is going on in Goldman's example of Henry and the barns. Henry and the physicist make judgments based on what is common in their past experience, while the mathematician is much more hesitant to postulate anything beyond the immediate content of her experience. This means that the mathematician is much less likely to ever make a false judgment. But it also means that she is much less likely to believe the full truth about the situation in front of her. To mistake a barn facade for a full barn, or to mistake a one-side-black sheep for a fully black sheep would be like being taken in by the Müller-Lyer illusion or the Ames room. But to insist on only believing in the facade and the one side of the sheep is ridiculous in ordinary contexts—and it would be even worse to constantly wonder if the room you just entered is actually shaped like an Ames room, with tiny or gigantic people at the other end. The disposition that leads Henry astray, and that the physicist endorses, and that all of us have on seeing people in new rooms, has unfortunate side effects in strange cases. But the theory of ecological rationality doesn't care about what goes on in those strange cases.

8.3.2 The "new evil demon"

Another relevant philosophical example is the "new evil demon" problem of Lehrer and Cohen (1983). The *old* evil demon problem is the one Descartes raised in his *Meditations on First Philosophy* (1641). Consider your belief that you are reading this book, in the location that you are currently experiencing. Can you be sure this belief is accurate? After all, if an all-powerful God is possible, then so is an all-powerful evil demon, and you could be some sort of being that is completely unlike a human, in a completely different environment from what you think, while the evil demon is deceiving you to believe that you are in this location, reading this book. This possibility aims to challenge our basic knowledge of what the world is like, and thus almost all of our knowledge (except perhaps the knowledge that I exist).

However, the *new* evil demon problem is about *justification*. One prominent theory of justification has been the idea that a belief is justified if and only if it is formed by a reliable process. However, it seems that, due to the influence of the demon, your counterpart in the demon world is forming their beliefs in a completely unreliable way. Thus, reliabilism says that their beliefs are unjustified. Lehrer and Cohen say this is unintuitive—we should think that the counterpart in the demon world is justified in the same way you are, even though their beliefs are unfortunately false.

Ecological rationality is in many ways like reliability. Both are concerned with beliefs being produced by a faculty that usually gets things right. However, in my account of ecological rationality, I want to emphasize that the range of circumstances in which your beliefs should do well depends on what environment you hope to be well adapted for. From our perspective in the actual world, the habits of the counterpart in the demon world are considered good ones, since they tend to get things right around here. But from the perspective of the demon world they might not be. Someone who was constantly hesitating about whether to believe she is in a physical world like ours or in a radically different world with an evil demon deceiver would be considered irrational around here. They would be very hesitant in either world. But they would go less wrong in one world than a person with habits that got them to the truth in the other world. To judge one of these dispositions as more ecologically rational than the other, we would need to say which range of worlds matter, and the relevant importance of getting the truth or avoiding incorrectness.

Legg and Hutter (2007) have argued that there is a particular probability distribution over environments (known elsewhere in the literature as the "Solomonoff prior") that is the appropriate one to consider when measuring the intelligence of an artificial intelligence, and thus that there is a single "universal intelligence" that is the unique objectively rational doxastic subject. Bostrom (2014) worries that an intelligence that is close enough to this one will be so effective at achieving its goals that, as long as its goals are not perfectly aligned with our own, we would face a grave risk from artificial intelligence. While I think that there should be concern about powerful unaligned artificial intelligences (just as there should be concern about powerful unaligned humans, corporations, governments, and even potentially animals, aliens, or angels), I don't think that there could be a unique appropriate probability distribution. In particular, this probability distribution will give some specific amount of weight to the kinds of environments that a creature faces on Earth in the 21st century, and some other amount of weight to the kinds of environments a creature would face on Alpha Centauri, and some other amount of weight to the kinds of environments a creature would face on Earth in the 12th century.

I posit that a creature that divides its credences in this way would be less worrisome to us than a creature that is optimized for only the variety of environments that tend to come up on Earth in the 21st century. This universal intelligence would do about as well in the evil demon world as it does in ours. It would be like the mathematician, who wonders whether the other side of the sheep is black until more evidence comes in.

8.3.3 The clairvoyant

One last case that is instructive to consider is that of Norman the Clairvoyant, in an example from BonJour (1980). In this case, Norman is a person who happens to have a clairvoyant ability. Whenever he instinctively considers the question of where the President is located, he happens to get it right. He has never considered whether he has this ability, and has never even thought about the question of whether he tends to get the location of the President right, or any of his other instinctive judgments. Today, the President happens to be in New York, and Norman happens to wonder about the location of the President, and thus instinctively forms the belief that the President is in New York.

BonJour's central point with this example is that from a reliabilist perspective, there is nothing wrong with Norman—he has a reliable faculty, and he has no evidence defeating his reliance on that faculty, so he is justified in his belief. But BonJour thinks this is unintuitive. He thinks Norman's belief is unjustified.

However, it seems to me that there are again multiple perspectives we can take. Within the range of nearby circumstances consisting of other possible locations of the President, Norman would form true judgments. If he had some practical or alethic reason to care about the location of the President, then trusting the beliefs formed through his clairvoyance would serve him well. But from a broader perspective, Norman is behaving badly. Trusting purely instinctual beliefs in general is in many cases a very bad guide to the truth. Over a broader range of propositions, Norman may well do better to suspend judgment on all instinctive beliefs for which he can't find a plausible justification, depending on the relative importance of the various subject matters on which he forms instinctive beliefs.

The reliabilist wants to say that beliefs formed via a reliable faculty are justified, even if the person has never considered the question of whether that faculty is reliable. One of their motivating concerns is that many people may not have considered the question of the reliability of their senses, and there doesn't appear to be any evidence outside of the senses that we could use to justify our belief in their reliability.

However, there are rules that we can follow that would improve our performance, and would make Norman give up on his beliefs. One policy involves trusting all instinctively formed beliefs. This policy would do well by Norman's Presidential beliefs, but would probably lead us all to form many bad judgments about other topics. Another policy involves trusting only the beliefs formed by the traditional five senses. This policy would do badly for Norman's Presidential beliefs, and would also do badly for various instinctive beliefs we form by other reliable means than the traditional five senses (for instance, beliefs formed using the less familiar proprioceptive or vestibular senses, as well as intuitive judgments about what our friends would do, or how the physical world around us would behave in ordinary cases). A more complicated policy would involve trusting judgments only when they seem to come from faculties that appear to be widely shared by people around us. This policy would likely approximate the judgments that BonJour has about Norman's case, and would cost us Norman's true beliefs, while saving us from the idiosyncratic false instinctive beliefs that most people have, and retaining the mixed bag of other intuitive beliefs. Most likely, most of us follow some policy that is much like this last policy, but with additional exceptions for cases of blurry vision, known illusions, and other further exceptions (though we probably haven't done much explicit consideration of the policy itself). There is likely an ecologically even better policy available to us, but likely not a whole lot better without making use of complicated math, science, and philosophy.

8.3.4 Conclusion of the case studies

In all these cases, I don't want to make a specific claim about whether the subjects involved are rational or not. This is partly because I don't have a theory of rationality for beliefs in the on-off sense, only for degrees of belief. But beyond that, I think there is also an important ambiguity in the concept of rationality. In some internal sense, rationality is a relatively weak requirement. All of the subjects considered (even the mathematician, physicist, and engineer) may count as rational in this weak sense, analogous to mere probabilism for degrees of belief. But in terms of ecological rationality, each will be more ecologically rational with respect to some ranges of environmental conditions, and less with respect to others. There is no objectively correct environment from which we have to evaluate ecological rationality. There are usually many. Furthermore, even considered from any particular relevant environment, it is unlikely that a person has the *optimal* policy.

Due to the fact that human evolution (both biological and cultural) occurred over tens of thousands of years in which humans moved between many climates and many cultural contexts, but were always on Earth and always living among other humans, we likely do quite well even in modern environments. But there is likely a policy that does even better in the particular set of environments under consideration.

Many people will have intuitions that agree with the following quote from Bruno de Finetti:

> If someone draws a house in perfect accordance with the laws of perspective, but choosing the most unnatural point of view, can I say that he is wrong? Well, this second example can be translated into an actual geometrical interpretation of the problem at hand. If we imagine an appropriate geometrical representation of the various logically possible cases, the various internally consistent opinions about their probabilities can analogously be considered as all the perspectives obtainable by varying our point of view. ...
>
> Anybody's taste will guide him to his best choice, and there might be cases where the aesthetic taste of most or even all people will agree more or less exactly on that choice. And

> in fact we see thousands and thousands of identical photographs, all showing some monument in the same perspective. But would it be legitimate—or, better, convenient—to interpret a free coincidence of tastes and opinions as the expression of an arcane metaphysical truth? (de Finetti, 1989, Section 6)

But as Frank Ramsey says, this is about pragmatic virtue—are the patterns useful or not? There is no abstract logical fact about whose perspective is right or wrong, just whose will actually be helpful for them.

> We all agree that a man who did not make inductions would be unreasonable: the question is only what this means. In my view it does not mean that the man would in any way sin against formal logic or formal probability; but that he had not got a very useful habit, without which he would be very much worse off, in the sense of being much less likely to have true opinions.
>
> This is a kind of pragmatism: we judge mental habits by whether they work, i.e. whether the opinions they lead to are for the most part true, or more often true than those which alternative habits would lead to. (Ramsey, 1926, p. 197)

Linguists (Pinker, 1994) and psychologists (Spelke and Kinzler, 2007) have argued that the human mind innately has the disposition to form certain types of judgments and learn certain types of patterns in the world around them. These dispositions go far beyond the content of experience, enabling infants to learn human languages, and to learn about gravity and social psychology, among many other topics, much faster than a truly "blank slate" mind that didn't have these mechanisms. If a human baby found itself in an environment that didn't include actual human languages, and didn't have physics and societies like those found in the real world, then the babies' mental development would lead them very far astray. But fortunately, most babies grow up in normal environments.

8.4 AN EVOLUTIONARY OBJECTION

As this discussion suggests, I think that the shaping of our doxastic dispositions by both cultural and biological evolution usually means that fairly ecologically rational doxastic dispositions will be quite common in humans. If so, then this can ground our judgment that a person who is quite confident that smoking does not cause lung cancer is irrational—the ecologically rational priors would allow existing evidence to make one quite confident that smoking causes lung cancer, and it's more likely that a person is failing to conditionalize than that they have the very extreme (and likely ecologically irrational) priors that would lead to that belief.

However, one might worry that a similar story would mean that *all* of our evolutionarily-favored cognitive dispositions count as ecologically rational. I have defended certain cognitive biases and heuristics as ecologically rational, even though they lead to false beliefs in well-known experimental circumstances. I have said that the reason we should expect many native human faculties to count as ecologically rational in this way is because these faculties (or faculties for developing dispositions like these in appropriate social environments) were shaped by natural selection. And yet it seems plausible that evolution (whether biological or cultural) could provide us with faculties that shouldn't count as rational. Mercier and Sperber (2019) argue that in fact our reasoning dispositions are quite irrational, and they were shaped to be so by our evolutionary history.

I don't have any expertise in evolutionary psychology, so I won't evaluate their arguments about the evolutionary history. And while I have more relevant expertise for considering questions about which contemporary habits of reasoning might in fact be good or bad in whatever sense one is interested in, I don't think it is productive to debate that question in this context either. Instead, I will just present a toy model in which a creature's evolutionary history could easily select for cognitive traits that are evolutionarily successful, but do not count as ecologically rational in the sense that I define.

The toy example I consider will be a kind of "testimonial sensibility", as Miranda Fricker (2007) puts it. She notes that traditional "reductionist" views in the epistemology of testimony seem to require people to do much more explicit reasoning about the reliability of others than we actually have time to do, and that the traditional rival "non-reductionist" views seem to advocate a problematic sort of gullibility. Instead, she says that we all have some sort of habitual subconscious way of judging the testimony we receive from others, likely partially based on experience, partially based on innate evolutionary dispositions, and possibly other sources as well. Much of her work focuses on the phenomenon of "testimonial injustice", whereby people in particular social roles get unfairly treated with either too much or too little credibility by their interlocutors. We seem to have a general tendency to accept the word of people in dominant social roles and high social status with great deference, while being quick to dismiss what is said by women, minorities, and other outgroups, particularly when it conflicts with what is said by people with high status.

I will show how such a disposition could conceivably be selected for evolutionarily, while still counting as ecologically irrational in my sense. I will work under the assumption that the role of a person's doxastic state is pragmatic, to guide them in successful action, and I think it will be easy to see that a similar argument should work if the role is alethic. On this pragmatic role for the doxastic state, a testimonial sensibility will be more ecologically rational to the extent that it helps the person act in ways that satisfy their desires more effectively. But importantly, what evolution selects for is slightly different from this—evolution will select for testimonial sensibilities that tend to lead to more copies of those testimonial sensibilities existing several generations later.

Importantly, the features of a testimonial sensibility that make it effective at leading to acts that satisfy one's desires are not necessarily the same as the features of a testimonial sensibility that make it likely to have more copies existing several generations later. To the extent that a person's desires are for things that help with their own survival and reproduction, and that of their close relatives, as well as their own sociocultural influence, there will be some alignment here. But to the extent that a person's desires are for other things, and to the extent that testimonial sensibilities can lead to survival and reproduction through means other than the promotion of actions, these will be misaligned.

A testimonial sensibility that includes some default tendency to defer to the beliefs of others on matters where you have little expertise and they have offered testimony (regardless of whether you have evidence of their honesty or expertise) is likely to have effects of many sorts. As long as the people who are speaking are in fact generally accurate in their beliefs (which over a wide range of topics is quite likely, since the people are in a position to communicate and themselves have belief-forming faculties that were selected through evolution) and also tend to be honest (which again is quite likely since most people one encounters are likely to be part of a broader community that selects for honesty), deferring to their beliefs is likely to lead one to the truth. It will also do so more effectively than a sensibility that insists on checking everything before believing what people say. This relatively deferential sensibility will enable accuracy, and also successful action, as judged by the desires and values of the person whose beliefs they are, and also survival and reproduction, to the extent that those desires and values support survival and reproduction.

But very clearly, not all of our desires and values are lined up with survival and reproduction. Some of us enjoy mountain climbing, and thus have desires and values that increase our risk of young death. Some of us are gay and prefer sexual activities that won't lead to reproduction over ones that might. Many of us develop deeper caring relationships with friends and associates who don't share our genes than with the people that do. And many of us have interests in solitary activities that make us less likely to have an impact on the broader social environment. Having more accurate beliefs would let us act even more effectively on the desires we actually have, which could sometimes be somewhat of a mixed bag from an evolutionary point of view. It is likely that enough of our desires require enough behavior that is correlated with biological flourishing for the self and one's family that there is some strong evolutionary pressure towards testimonial sensibilities that lead to true beliefs (and effective acts). But this selective pressure is weaker than it would be if all of our desires were perfectly aligned with evolutionary fitness (whether biological or cultural).

Our testimonial sensibilities also have effects on our behavior and on our evolutionary fitness through means other than the way they lead to beliefs that guide our actions. Van Leeuwen and Lombrozo (2023) note that many of our beliefs have the effect of marking our membership in social groups, beyond whatever actions they might guide, and whatever accuracy they might or might not have. To the extent that membership in different social groups is relevant for survival, reproduction, and cultural impact, evolution will select for testimonial sensibilities that promote beliefs that get us into the right social groups. And this is where I think selection for testimonial injustice is likely to come about.

Sharing beliefs with people of high social status will likely lead to membership in powerful social groups. Sharing beliefs with outsiders and people of low social status will likely not as often be as helpful in this way. When there is no intergroup conflict and when beliefs are shared across social groups, this effect is likely not that relevant. But when one receives disagreeing testimony from individuals of very different social status, it may end up being quite evolutionarily relevant to prefer to affiliate with the high-status person than the low-status person. Many "standpoint epistemologists" (Anderson, 2024) point out that in fact, there are many cases in which outsiders and low-status members of a society often have more accurate beliefs, particularly about existing social injustices. But if accurate beliefs about these facts tend to lead to acts that are more effective at promoting personal desires rather than promoting survival, reproduction, and broader cultural impact, then this will be the kind of accuracy that evolution doesn't particularly select for.

Thus, my claim is that most innate human doxastic dispositions will, because of their selection through evolutionary history, be ones that promote survival and reproduction for oneself and close family, and social influence on others. To the extent that many human desires tend to require survival and lead to reproductive and cultural success, this will often mean that these doxastic dispositions tend to be relatively ecologically rational. But there are some predictable ways in which beliefs have evolutionary effects through means other than the acts they guide (or their accuracy), and the kinds of dispositions that are most relevant to these may tend to be more irrational.

Note that my response here distinguishes my view from one offered by Susanna Rinard (2017). Rinard argues that belief should be treated purely pragmatically, like any other sort of behavior that can lead to the satisfaction of one's desires. She judges a belief as rational to the extent that it helps one achieve what one wants, whether it is by guiding action or by being helpful in some other way. I claim instead that, even if belief is judged through its practical consequences, there is a particular role belief plays—it is only to the extent that belief *guides successful action* that it is said to be rational. If it gives rise to good results in other ways (through promoting one's social status, or through other means, like Pascal's Wager, or the "epistemic bribes" discussed in Section 3.4.1), this is not relevant to its rationality as belief.

This discussion is not meant to be a thorough or accurate discussion of the origin of testimonial sensibilities. It is meant as a "just-so" story that shows how things *could* work, so that evolution would favor many cognitive faculties that do contribute to a high level of ecological rationality, but also some cognitive faculties that do not. For more detailed work on how biological and especially cultural evolution could give rise to human faculties that do well (in one way or another) in home environments, see Henrich (2016). Importantly, he shows that many of the traditional ways that humans gather knowledge about the world, through deference to community elders and to members of one's own gender that are slightly older than oneself, actually do end up enabling more reliably accurate and effective beliefs than the Enlightenment-approved methods of questioning authority and thinking for oneself. Levy and Alfano (2020) argue that this demonstrates the possibility of a kind of knowledge through epistemic vice—but I suggest that we might be more careful about how to determine which psychological traits constitute virtues and which constitute vices.

8.5 IDEAL AND NON-IDEAL

In Chapter 7, I suggested that there are at least three very different sorts of norms that apply to epistemic agents—norms of logical omniscience that ought to shape their set of doxastic possibilities, norms of probabilism that ought to shape their degrees of belief given their doxastic possibilities, and diachronic norms for how to update over time. In this chapter, I have suggested that there is a natural externalist strengthening of some of these, in the form of the concept of "ecological rationality".

Much application of Bayesian epistemology involves studying how ideal subjects behave. We assume that subjects are perfectly logically omniscient and always update perfectly, and have priors that are ecologically rational for the sort of problem they are considering in the environment they are in. (The subjects in these examples are often perfectly certain that the coins are fair and the decks have 52 cards, and that the stipulations of the thought experiment are correct.) Then we derive some consequences of these ideas, and use them to characterize some of the features of perfectly ideal reasoners.

However, actual agents will not be ideal in these ways. Regardless of whether we are considering humans or animals or scientific communities or any other sort of entity to which Bayesianism is meant to apply, these entities will not perfectly match the recommendations of the theory. But there's no obvious way to compare the types of failure. If one person has degrees of belief 0.51 in both a proposition and its complement, while another has degree of belief 0.9 in the proposition and 0.1 in its complement, the latter may satisfy the internalist norms perfectly while the former does not. However, the former may be doing far better by the externalist ecological norms.

Herb Simon says, "Human rational behavior (and the rational behavior of all physical symbol systems) is shaped by a scissors whose two blades are the structure of the task environments and the computational capabilities of the actor" (Simon, 1990, p. 7). This helps us understand the force of ecological norms. Philosophers have a tendency to emphasize the computational capabilities of the actor, and focus on norms of ideal logical omniscience, and perfect Bayesian probabilism and conditionalization, because a system that can do these things can avoid certain problems that come up in any environment. But no system actually has the capacity to ensure perfect compliance with these norms. They will instead often do better by trying to push against the features of their environment in ways that fix the problems that are most pressing, and even to take advantage of reliable environmental regularities to quickly compute things that they can't compute as easily inside themselves.

On either the pragmatic or alethic conception of credence, having maximal degree of belief in every truth and minimal degree of belief in every falsehood is better than any other credal state. However, for limited physical beings, any set of policies that pushed us to these degrees of belief would also push us to these same degrees of belief in other possible worlds where the set of truths and falsehoods is quite different. Thus, even though these degrees of belief would be better than the ones we have, they would be produced by dispositions that would not be valuable across the whole range of ecological possibilities.

Even for doxastic subjects that are not themselves groups, we can think of the set of selves distributed across these nearby worlds as forming a kind of political community. Because of the features of the system in which they are bound, these selves must operate under a single set of cognitive policies to guide their cognitive lives in their different worlds. This policy can have some sensitivity to the differences between the worlds in which they live, but no system has perfect sensitivity to all these differences. Therefore, they must agree on a policy that will inevitably lead most of them to suboptimal possibilities. Policies that work well for one may not work well for another. A policy that leads all of them to have high credences in many false propositions may be bad. But a policy that prevents this, like a policy of always having credence 0.5 in everything, may well be worse. A policy that leads some of them to non-probabilistic degrees of belief is going to lead to problems of a sort that could be detected. But a policy of paying attention to those kinds of issues will sometimes take away attention from more pressing issues.

Some think of the role of a theory of ideal rationality as providing a goal that non-ideal subjects can aim towards. But I don't think this is right. I think that *success* is a goal that *everyone*, both ideal and non-ideal subjects, should aim towards.

Ideal subjects find some optimal way of trading off the different possibilities of success against each other. This might be the unique way of trading off costs and benefits across a particular set of ecologically-relevant possibilities. Or it might be one of the many internally coherent ways of trading off costs and benefits in whatever states the ideal subject deems possible. For any such ideal subject, there is no way of getting some of the selves distributed across these possibilities any more success without costing success for others.

On one way of thinking of non-ideal subjects, there are options they can do with no new information that would lead all of their members to greater success. By moving to the closest probabilistic doxastic state to the agent's own, they guarantee that all of their members get closer to the truth. But figuring out precisely what move this is can be quite costly, and often much more difficult than just making one more observation of the world around them. If they care about ideality, then they should spend the effort to figure out what this closest probabilistic state is, rather than making a new observation and updating in some non-ideal way. But I claim that they should only care about ideality instrumentally, to the extent that it is a way to get at what actually matters, which is success. If the new observation can bring them much closer to the truth, that is likely more valuable than fiddling around inside their head. (Something like these observations is, I think driving some of the ideas of Staffel (2017), Singer and Aronowitz (2022), and Daoust (2024).)

Bayesianism in its abstract sense is a theory of ideal subjects. The ideal doxastic subjects (whether pragmatic or alethic) have degrees of belief that are well-represented by a numerical function that satisfies the probability axioms, and they update it in line with conditionalization. To the extent that non-ideal subjects are close to ideal, they will behave in similar ways. So Bayesianism can help us understand how non-ideal subjects will behave, provided they are relatively close to ideal. In the kinds of simple thought experiments often considered by philosophers (including things like Sleeping Beauty (Elga, 2000), and Doomsday (Leslie, 1990)), and many of the simple experimental setups of economists and psychologists, this is good enough. Understanding how an ideal subject behaves in these cases is good enough to understand the rough ideas of how a non-ideal subject will behave.

But when it comes to questions of how to deal with expected irrationality (including in cases of peer disagreement (Elga, 2007), higher-order evidence of irrationality (Christensen, 2010), and perhaps the cases of credence-state correlation discussed in Section 3.4.1), Bayesianism may be less helpful. And when it comes to building artificial agents for extremely large and complex tasks, like scientific communities

(Kitcher, 1990), democratic societies (Anderson, 2006), and computer models of language (Vaswani et al., 2017), we don't try to explicitly build in the Bayesian apparatus. It's likely that parts of these complex systems will behave in roughly Bayesian ways (as the output neurons of a neural net do). But we shouldn't expect that explicitly building in Bayesian reasoning will help these systems perform better.

There is another way of thinking of non-ideal rationality, that tries to make explicit the costs and benefits of the deliberation process itself, in addition to thinking about the possibilities and the outcomes. One can sometimes identify an optimal algorithm for a particular environment, subject to these sorts of costs and benefits. A creature who is actually in this situation wouldn't be able to spare the resources to figure out what is optimal—but if it is lucky (or if it is well-designed by an intentional designer, or well-evolved by a long process of cultural and social evolution) it may behave close to optimally. We might think of this as a kind of "ideal non-ideality". I take it that this is the project of Icard (2026) and Thorstad (2024). For any such situation, Bayesianism will likely emerge as a limiting solution, as the costs of deliberation become negligible.

8.6 CONCLUSION

Mona, the monarch butterfly, is born in Missouri in August. Her mother was born in June in Minnesota, and her grandmother was born there in April. Her great-grandmother was born in February in Texas. A few weeks after she is born, when even Missouri was starting to get cooler, Mona starts flying southwards, through her great-grandmother's homeland, where she has never been. Along the way she eats and builds up fat stores. In December, she finally reaches reaches a grove of trees in Mexico that have the right humidity to preserve her without eating, and she spends two months there in a semi-hibernating state, occasionally flapping around. In February, once the days are 11 hours long again, she starts flying northwards. In Texas, she finally lays her eggs and soon dies. The next three generations behave as her great-grandmother, grandmother, and mother did—the first generation lives a short life flying northward through the prairies, to a northern homeland where the second generation lives its life. The third generation starts moving south again. But at the fourth generation, again the next butterfly eats more in fall, flies far south and lives for a few months without eating, to a place no one in her family has been for four generations.

We don't know what the mental life of butterflies is, to the extent that they even have mental lives. But somehow they manage to guide themselves southwards and northwards to locations they have never been as individuals. While it is conceivable that there is some sort of cultural communication in the butterflies, it seems more likely that the information that guides her to the particular forests where she can survive in the mountains of Mexico is somehow instinctive. Whatever this sense of direction is, it seems to be mostly reliable. Most butterflies in North America with these habits do quite well at achieving whatever sort of goals they have.

However, if these butterflies were transported to Australia, or Central Asia, or to a more alien environment with different climate patterns, they probably wouldn't do all that well. A strategy that is "rational" in the sense of doing well from year to year and decade to decade and millennium to millennium in North America may not work well in a broader environment.

It may well be that there are many constraints beyond mere probabilism that are relevant to our dramatic success in the world as we know it. We tend to have broad agreement about the sorts of evidence that confirms various scientific theories—it seems odd if someone takes the existence of fossil dinosaurs to disconfirm the theory of evolution. But I claim that someone with those sorts of dispositions could be internally just as rational as the rest of us. It is only from an external perspective that these probabilistic dispositions can be distinguished, and this external perspective will be privileged only because of some feature of our environment, not by rationality itself. Similarly, we are familiar with many sorts of illusions, both optical illusions (like the waterfall illusion) and cognitive illusions (like availability bias and anchoring). It may well be that these features of our epistemic system are counterproductive in many environments, including some of the modern environments that we often find ourselves in. But they may well be rational from the perspective of the range of circumstances that humans have historically typically found themselves in.

This leads us to a partial explanation of the sub-title of this book. A Bayesian agent must have some probability function as their degrees of belief, but there is nothing that can be agreed upon by all conceivable Bayesian agents that would say which function is right. Any Bayesian agent must be opinionated in one way or another in order to function. Sometimes, these opinions will be wildly successful beyond anything we can imagine, as in the case of the butterfly. But these opinions will not always be shared. When they are shared, there can be greater co-operation, but one can learn from and treat with respect other agents with other opinions. My introduction to the philosophical foundations of Bayesianism follows from my own opinionated views of where current research into these foundations has left us. Other researchers may see the same evidence and come to different conclusions. None of us is likely to get everything right. But hopefully we can all learn from each other.

Research Questions

1. Are there pragmatic or alethic motivations for any theory of rationality other than radically subjective Bayesianism, ecological rationality, or omniscience?

2. Can ecological rationality explain the idea of justification of belief?

3. Can ecological rationality explain all the case judgments of moderate permissivists and uniqueness theorists about credence?

4. Given the kinds of non-ideality of actual doxastic subjects (whether human or others), what kinds of theories better describe them than Bayesian theory?

5. Do the kinds of non-ideality of actual doxastic subjects mean that they are best understood in some way other than the pragmatic or alethic ways that I have been proposing throughout?

Bibliography

Adams, E. (1965). The logic of conditionals. *Inquiry*, 8:166–197.

Ahmed, A. (2013). Causal decision theory: A counterexample. *Philosophical Review*, 122(2):289–306.

Ajalloeian, A. (2025). Probabilism through a transform: An improved accuracy argument for probabilism. *Philosophy and Phenomenological Research*, 111(3):1159–1186.

Alchourròn, C. E., Gärdenfors, P., and Makinson, D. (1985). On the logic of theory change: Partial meet contraction and revision functions. *Journal of Symbolic Logic*, 50:510–530.

Allais, M. (1953). Le comportement de l'homme rationnel devant le risque: Critique des postulats et axioms de l'école Americaine. *Econometrica*, 21:503–546.

Anderson, E. (2006). The epistemology of democracy. *Episteme*, 3(1):8–22.

Anderson, E. (2024). Feminist epistemology and philosophy of science. *Stanford Encyclopedia of Philosophy*.

Anscombe, G. E. M. (1963). *Intention*. Blackwell.

Arló-Costa, H. and Pedersen, A. P. (2012). Belief and probability: A general theory of probability cores. *International Journal of Approximate Reasoning*, 53(3):293–315.

Armendt, B. (1986). A foundation for causal decision theory. *Topoi*, 5:3–19.

Arntzenius, F. (2003). Some problems for conditionalization and reflection. *The Journal of Philosophy*, 100(7):356–371.

Axelrod, R. (1980). Effective choice in the Prisoner's Dilemma. *Journal of Conflict Resolution*, 24(1):3–25.

Bacon, A. (2015). Stalnaker's thesis in context. *The Review of Symbolic Logic*, 8(1):131–163.

Barcan, R. (1946). A functional calculus of first order based on strict implication. *The Journal of Symbolic Logic*, 11(1):1–16.

Basu, R. (2018). The wrongs of racist beliefs. *Philosophical Studies*, 176(9):2497–2515.

Basu, R. (2023a). Morality of belief I: How beliefs wrong. *Philosophy Compass*, 18(7):1–10.

Basu, R. (2023b). Morality of belief II: Three challenges and an extension. *Philosophy Compass*, 18(7):11–19.

Bayes, T. (1763). An essay towards solving a problem in the doctrine of chances. *Philosophical Transactions of the Royal Society*, 53:370–418.

Bender, E. and Koller, A. (2020). Climbing towards NLU: On meaning, form, and understanding in the age of data. In *Proceedings of the Annual Meeting of the Association for Computational Linguistics*, volume 58, pages 5185–5198.

Bennett, J. (2003). *A Philosophical Guide to Conditionals*. Oxford.

Berker, S. (2013). Epistemic teleology and the separateness of propositions. *Philosophical Review*, 122(3):337–393.

Berto, F. and Jago, M. (2018). Impossible worlds. *Stanford Encyclopedia of Philosophy*.

Blackwell, K. and Drucker, D. (2019). When propriety is improper. *Philosophical Studies*, 176(2):367–386.

Boespflug, Mark and Jackson, E. (2024). Doxastic voluntarism. *Stanford Encyclopedia of Philosophy*.

BonJour, L. (1980). Externalist theories of empirical knowledge. *Midwest Studies in Philosophy*, 5(1):53–73.

Bostrom, N. (2014). *Superintelligence: Paths, Dangers, Strategies*. Oxford University Press.

Bradley, D. (2015). *A Critical Introduction to Formal Epistemology*. Routledge.

Brennan, P., Crispo, A., Zaridze, D., Szeszenia-Dabrowska, N., Rudnai, P., Lissowska, J., Fabiánová, E., Mates, D., Bencko, V., Foretova, L., Janout, V., Fletcher, T., and Boffetta, P. (2006). High cumulative risk of lung cancer death among smokers and nonsmokers in central and eastern europe. *American Journal of Epidemiology*, 164(12):1233–1241.

Brier, G. (1950). Verification of forecasts expressed in terms of probability. *Monthly Weather Review*, 78(1):1–3.

Briggs, R. (2009). Distorted reflection. *Philosophical Review*, 118(1):59–85.

Briggs, R. and Pettigrew, R. (2016). An accuracy-dominance argument for conditionalization. unpublished.

Broome, J. (2007). Wide or narrow scope? *Mind*, 116(462):359–370.

Buchak, L. (2012). Can it be rational to have faith? In Chandler, J. and Harrison, V., editors, *Probability in the Philosophy of Religion*. Oxford University Press.

Buchak, L. (2013). *Risk and Rationality*. Oxford University Press.

Buchak, L. (2014). Belief, credence, and norms. *Philosophical Studies*, 169(2):285–311.

Buchak, L. (2017). Taking risks behind the veil of ignorance. *Ethics*, 127(3):610–644.

Buchanan, R. and Dogramaci, S. (2024). Belief about probability. *Journal of Philosophy*. https://philarchive.org/rec/BUCBAP

Burgess, A., Cappelen, H., and Plunkett, D., editors (2020). *Conceptual Engineering and Conceptual Ethics*. Oxford University Press.

Butler, J. (1736). *The Analogy of Religion to the Constitution and Course of Nature*. James, John & Paul Knapton.

Caie, M. (2013). Rational probabilistic incoherence. *The Philosophical Review*, 122(4).

Caie, M. (2020). A problem for credal consequentialism. In Dunn, J. and Ahlstrom-Vij, K., editors, *Epistemic Consequentialism*. Oxford University Press.

Carr, J. (2017). Epistemic utility theory and the aim of belief. *Philosophy and Phenomenological Research*, 95(3):511–534.

Carr, J. (2021). A modesty proposal. *Synthese*, 198(4):3581–3601.

Chalmers, D. (2012). *Constructing the World*. Oxford.

Christensen, D. (1991). Clever bookies and coherent beliefs. *The Philosophical Review*, 100(2):229–247.

Christensen, D. (1996). Dutch book arguments de-pragmatized: Epistemic consistency for partial believers. *The Journal of Philosophy*, 93(9):450–479.

Christensen, D. (1999). Measuring confirmation. *The Journal of Philosophy*, 96(9):437–461.

Christensen, D. (2001). Preference-based arguments for probabilism. *Philosophy of Science*, 68(3):356–376.

Christensen, D. (2010). Rational reflection. *Philosophical Perspectives*, 24:121–140.

Churchland, P. (1981). Eliminative materialism and the propositional attitudes. *Journal of Philosophy*, 78:67–90.

Churchland, P. (1986). *Neurophilosophy: Toward a Unified Science of the Mind/Brain*. MIT Press.

Clark, A. and Chalmers, D. (1998). The extended mind. *Analysis*, 58:10–23.

Clifford, W. K. (1877). The ethics of belief. *Contemporary Review*, 29:289–309.

Coady, C. A. J. (1990). *Testimony: A Philosophical Study*. Clarendon Press.

Mollo and Millière (2023). The vector grounding problem. Technical report, Macquarie University.

Colmer, J. (2020). What is the meaning of (statistical) life? benefit-cost analysis in the time of COVID-19. *Oxford review of economic policy.*

Cox, R. T. (1946). Probability, frequency and reasonable expectation. *American Journal of Physics*, 14(1):1–13.

Dallmann, J. (2015). *Belief as a Credal Plan.* PhD thesis, University of Southern California.

Dallmann, J. (2017). When obstinacy is a better (cognitive) policy. *Philosophers' Imprint*, 17(24):1–18.

Danzig, R. (2022). Machines, bureaucracies, and markets as artificial intelligences. Technical report, Center for Security and Emerging Technology.

Daoust, M.-K. (2024). What second-best epistemology could be. *Analytic Philosophy.*

Daoust, M.-K. and Côté-Bouchard, C. (2023). Epistemic consequentialism, veritism, and scoring rules. *Erkenntnis*, 88:1741–1765.

Das, N. (2022). Externalism and exploitability. *Philosophy and Phenomenological Research*, 104(1):101–128.

de Finetti, B. (1931). Probabilismo. *Logos*, 14:163–219.

de Finetti, B. (1937). La prévision: ses lois logiques, ses sources subjectives. *Annales de l'Institut Henri Poincaré*, 7:1–68.

de Finetti, B. (1964). Foresight: Its logical laws, its subjective sources. In Kyburg, H. and Smokler, H., editors, *Studies in subjective probability*, pages 94–158. Wiley.

de Finetti, B. (1974). *Theory of Probability*, volume 1. Wiley.

de Finetti, B. (1989). Probabilism. *Erkenntnis*, 31:169–223.

Descartes, R. (1641). *Meditations on First Philosophy.*

Diaconis, P. and Zabell, S. (1982). Updating subjective probability. *Journal of the American Statistical Association*, 77(380):822–830.

Dorst, K. (2019). Lockeans maximize expected accuracy. *Mind*, 128(509):175–211.

Dorst, K., Levinstein, B., Salow, B., Husic, B., and Fitelson, B. (2021). Deference done better. *Philosophical Perspectives*, 35(1):99–150.

Dorst, K. and Mandelkern, M. (2021). Good guesses. *Philosophy and Phenomenological Research*, 105(3):581–618.

Douven, I. (2013). Inference to the best explanation, Dutch books, and inaccuracy minimisation. *The Philosophical Quarterly*, 63(252):428–444.

Douven, I. and Williamson, T. (2006). Generalizing the lottery paradox. *The British Journal for the Philosophy of Science*, 57(4):755–779.

Dreier, J. (1996). Rational preference: Decision theory as a theory of practical rationality. *Theory and Decision*, 40(3):249–276.

Dretske, F. (1986). Misrepresentation. In Bogdan, R., editor, *Belief: Form, Content, and Function*. Oxford University Press.

Dreyfus, H. and Dreyfus, S. (1988). *Mind over Machine*. Free Press.

Earman, J. (1992). *Bayes or Bust?* The MIT Press.

Easwaran, K. (2008). *The Foundations of Conditional Probability*. PhD thesis, University of California, Berkeley.

Easwaran, K. (2013a). Expected accuracy supports conditionalization — and conglomerability and reflection. *Philosophy of Science*, 80(1):119–142.

Easwaran, K. (2013b). Why countable additivity? *Thought*, 1(4):53–61.

Easwaran, K. (2014a). Decision theory without representation theorems. *Philosophers' Imprint*, 14(27):1–30.

Easwaran, K. (2014b). Regularity and hyperreal credences. *The Philosophical Review*, 123(1):1–41.

Easwaran, K. (2015). Dr. Truthlove, or, how I learned to stop worrying and love Bayesian probability. *Noûs*.

Easwaran, K. (2017). The tripartite role of belief: Evidence, truth, and action. *Res Philosophica*, 94(2):189–206.

Easwaran, K. (2019). Conditional probability. In Pettigrew, R. and Weisberg, J., editors, *Open Handbook of Formal Epistemology*, chapter 4, pages 131–198. PhilPapers.

Easwaran, K. (2021). A classification of Newcomb problems and decision theories. *Synthese*, 198:6415–6434.

Easwaran, K. (2025). Clarifications on countable additivity. https://philpapers.org/rec/EASCOC.

Easwaran, K. (2026). Generalizations of risk-weighted expected utility. *Economics and Philosophy*.

Easwaran, K. and Fitelson, B. (2012). An 'evidentialist' worry for Joyce's argument for probabilism. *Dialectica*, 66(3):425–433.

Easwaran, K., Levinstein, B., and Shear, T. (2025). Tickles, iteration, and habits. *Theory and Decision*.

Easwaran, K. and Nielsen, M. (2025). Updating by maximizing expected accuracy in infinite, non-partitional settings. *Journal of Philosophical Logic*, 54, 1095–1134.

Easwaran, K. and Stern, R. (2026). Diachronic and interpersonal coherence. In Reed, B. and Flowerree, A., editors, *Towards an Expansive Epistemology: Norms, Action, and the Social Sphere*.

Edgington, D. (1995). On conditionals. *Mind*, 104(414):235–329.

Egan, A. (2007). Some counterexamples to causal decision theory. *Philosophical Review*, 116(1):93–114.

Elga, A. (2000). Self-locating belief and the Sleeping Beauty problem. *Analysis*, 60(2):143–147.

Elga, A. (2007). Reflection and disagreement. *Noûs*, 41(3):478–502.

Elga, A. (2010). Subjective probabilities should be sharp. *Philosophers' Imprint*, 10(5).

Elga, A. and Rayo, A. (2021a). Fragmentation and information access. In Borgoni, C., Kinderman, D., and Onofri, A., editors, *The Fragmented Mind*, pages 37–53. Oxford.

Elga, A. and Rayo, A. (2021b). Fragmentation and logical omniscience. *Noûs*, 56(3):716–741.

Eriksson, L. and Hájek, A. (2007). What are degrees of belief? *Studia Logica*, 86(2):183–213.

Evnine, S. (2003). Epistemic unities. *Erkenntnis*, 59:365–388.

Evnine, S. (2005). Containing multitudes: Reflection, expertise, and persons as groups. *Episteme*, 2(1):57–64.

Fantl, J. and McGrath, M. (2002). Evidence, pragmatics, and justification. *Philosophical Review*, 111(1):67–94.

Feldman, R. (2006). Epistemological puzzles about disagreement. In Hetherington, S., editor, *Epistemology Futures*, pages 209–231. New York: Oxford University Press.

Field, H. (1978). A note on Jeffrey conditionalization. *Philosophy of Science*, 45(3):361–367.

Figdor, C. (2018). *Pieces of Mind.* Oxford University Press.

Fisher, R. (1959). *Smoking The Cancer Controversy: Some Attempts to Assess the Evidence.* Oliver and Boyd.

Fitelson, B. (1999). The plurality of Bayesian measures of confirmation and the problem of measure sensitivity. *Philosophy of Science*, 66(3):S362–S378.

Fodor, J. (1975). *The Language of Thought.* Thomas Y. Crowell.

Foley, R. (1993). *Working Without a Net: A Study of Egocentric Epistemology.* Oxford.

Frege, G. (1892). Über Sinn und Bedeutung. *Zeitschrift für Philosophie und philosophische Kritik.*

Fricker, M. (2007). *Epistemic Injustice: Power and the ethics of knowing.* Oxford University Press.

Frost-Arnold, K. (2016). Social media, trust, and the epistemology of prejudice. *Social Epistemology*, 30(5-6):513–531.

Fusco, M. (2025). Imaging and the diachronic Dutch book. Philosophy And Phenomenological Research, 1–33. https://doi.org/10.1111/phpr.70065.

Gallow, J. D. (2017). No one can serve two epistemic masters. *Philosophical Studies*, 175:2389–2398.

Gallow, J. D. (2021). Updating for externalists. *Noûs*, 55(3):487–516.

García-Garibay, O. B. and de Lafuente, V. (2015). The Müller-Lyer illusion as seen by an artificial neural network. *Frontiers in Computational Neuroscience*, 9(21).

Gardiner, G. (2018). Evidentialism and moral encroachment. In McCain, K., editor, *Believing in Accordance with the Evidence: New Essays on Evidentialism*, pages 169–195. Springer-Verlag.

Gelman, A. and Nolan, D. (2002). You can load a die, but you can't bias a coin. *The American Statistician*, 56(4):308–311.

Gelman, A. and Shalizi, C. R. (2013). Philosophy and the practice of Bayesian statistics. *British Journal of Mathematical and Statistical Psychology*, 66:8–38.

Gendler, T. (2008). Alief and belief. *The Journal of Philosophy*, 105(10):634–663.

Gettier, E. (1963). Is justified true belief knowledge? *Analysis*, 23(6):121–123.

Gigerenzer, G. and Todd, P. M. (1999). Ecological rationality: the normative study of heuristics. In *Ecological rationality: Intelligence in the world*. Oxford University Press.

Gilbert, M. (1987). Modelling collective belief. *Synthese*, 73(1):185–204.

Gilboa, I., Postlewaite, A., and Schmeidler, D. (2012). Rationality of belief or: why Savage's axioms are neither necessary nor sufficient for rationality. *Synthese*, 187:11–31.

Goldman, A. (1976). Discrimination and perceptual knowledge. *The Journal of Philosophy*, 73(20):771–791.

Goldman, A. (1979). What is justified belief? In Pappas, G. S., editor, *Justification and Knowledge*, pages 1–25. Reidel.

Goldstein, S. and Levinstein, B. (2024). Does ChatGPT have a mind? ArXiV.

Good, I. J. (1967). On the principle of total evidence. *British Journal for the Philosophy of Science*, 17(4):319–321.

Good, I. J. (1985). Weight of evidence: a brief survey. In Bernardo, J., DeGroot, M., Lindley, D., and Smith, A., editors, *Bayesian Statistics 2*, pages 249–270. Elsevier.

Goodman, N. (1955). The new riddle of induction. In *Fact, Fiction, and Forecast*. Harvard University Press.

Goodman, N., Tenenbaum, J., and Gerstenberg, T. (2015). Concepts in a probabilistic language of thought. In Margolis, E. and Laurence, S., editors, *The Conceptual Mind: New Directions in the Study of Concepts*. MIT Press.

Greaves, H. (2013). Epistemic decision theory. *Mind*, 122(488):915–952.

Greaves, H. and Wallace, D. (2006). Justifying conditionalization: Conditionalization maximizes expected epistemic utility. *Mind*, 115(459):607–632.

Hacking, I. (1967a). Slightly more realistic personal probability. *Philosophy of Science*, 34(4):311–325.

Hacking, I. (1967b). Slightly more realistic personal probability. *Philosophy of Science*, 34(4):311–325.

Hájek, A. (2003). What conditional probability could not be. *Synthese*, 137:273–323.

Hájek, A. (2016). Deliberation welcomes prediction. *Episteme*, 13(4):507–528.

Hájek, A. and Smithson, M. (2011). Rationality and indeterminate probabilities. *Synthese*, 187:33–48.

Halpern, J. (1999a). A counterexample to theorems of Cox and Fine. *Journal of Artificial Intelligence Research*, 10:76–85.

Halpern, J. (1999b). Cox's theorem revisited. *Journal of Artificial Intelligence Research*, 11:429–435.

Halpern, J. (2009). Lexicographic probability, conditional probability, and nonstandard probability. *Games and Economic Behavior.*

Halpern, J. (2017). *Reasoning about Uncertainty.* MIT Press.

Harman, G. (1986). *Change in View: Principles of Reasoning.* MIT Press.

Haslanger, S. (2012). *Resisting Reality.* Oxford University Press.

Hawthorne, J., Rothschild, D., and Spectre, L. (2016). Belief is weak. *Philosophical Studies*, 173(5):1393–1404.

Hedden, B. (2015). *Reasons without Persons.* Oxford University Press.

Hempel, C. (1945). Studies in the logic of confirmation (i.). *Mind*, 54(213):1–26.

Hempel, C. (1962). Deductive-nomological vs. statistical explanation. In Feigl, H. and Maxwell, G., editors, *Minnesota Studies in the Philosophy of Science*, volume 3, pages 98–169. University of Minnesota Press.

Henrich, J. (2016). *The Secret of Our Success.* Princeton University Press.

Hieronymi, P. (2006). Controlling attitudes. *Pacific Philosophical Quarterly*, 87(1):45–74.

Hilbert, D. (1899). *Grundlagen der Geometrie.* Teubner.

Hilbert, D. (1902). *The Foundations of Geometry.* Open Court.

Hild, M. (1998). Auto-epistemology and updating. *Philosophical Studies*, 92(3):321–361.

Hintikka, J. (1962). *Knowledge and Belief: an introduction to the logic of the two notions.* College Publications.

Holguín, B. (2022). Thinking, guessing, and believing. *Philosophers' Imprint*, 22(1):1–34.

Holton, R. (2015). Intention as a model for belief. In Vargas, M. and Yaffe, G., editors, *Rational and social agency: Essays on the philosophy of Michael Bratman.* Oxford University Press.

Horowitz, S. (2013). Immoderately rational. *Philosophical Studies*, 167(1):1–16.

Horowitz, S. and Dogramaci, S. (2016). Uniqueness: a new argument. *Philosophical Issues*, 26.

Hosiasson-Lindenbaum, J. (1940). On confirmation. *Journal of Symbolic Logic*, 5(4):133–148.

Howe, C. and Purves, D. (2004). The Müller-Lyer illusion explained by the statistics of image-source relationships. *Proceedings of the National Academy of Sciences*, 102(4):1234–1239.

Howson, C. and Urbach, P. (1989). *Scientific Reasoning: The Bayesian Approach.* Open Court.

Icard, T. (2026). *Resource Rationality.* MIT Press.

Ichikawa, J. J. and Steup, M. (2017). The analysis of knowledge. *Stanford Encyclopedia of Philosophy.*

Isaacs, Y. and Russell, J. (2023). Updating without evidence. *Noûs.*

Jackson, E. (2019). *Belief and Credence: A Defense of Dualism.* PhD thesis, University of Notre Dame.

Jackson, E. (2020). The relationship between belief and credence. *Philosophy Compass*, 15(6):1–13.

Jackson, E. (2022). Why credences are not beliefs. *Australasian Journal of Philosophy*, 100(2):360–370.

James, W. (1896). The will to believe. *The New World*, 5:327–347.

Jaynes, E. T. (2003). *Probability Theory: The Logic of Science.* Cambridge University Press.

Jeffrey, R. (1965). *The Logic of Decision.* McGraw-Hill.

Jeffrey, R. (1970). Dracula meets wolfman: Acceptance vs. partial belief. In Swain, M., editor, *Induction, Acceptance, and Rational Belief.* Dordrecht: Reidel.

Jeffrey, R. (1984). Bayesianism with a human face. In Earman, J., editor, *Testing Scientific Theories*, volume 10 of *Minnesota Studies in the Philosophy of Science*, pages 133–156. University of Minnesota Press.

Jorgensen Bolinger, R. (2020). Varieties of moral encroachment. *Philosophical Perspectives*, 34(1):5–26.

Joyce, J. (1998). A nonpragmatic vindication of probabilism. *Philosophy of Science*, 65(4):575–603.

Joyce, J. (1999). *The Foundations of Causal Decision Theory*. Cambridge University Press.

Joyce, J. (2005). How probabilities reflect evidence. *Philosophical Perspectives*, 19:153–178.

Joyce, J. (2009). Accuracy and coherence: Prospects for an alethic epistemology of partial belief. In *Degrees of Belief*, volume 342 of *Synthese Library*, pages 263–297. Springer.

Kahneman, D. (2011). *Thinking, Fast and Slow*. Farrar, Straus and Giroux.

Kahneman, D. and Tversky, A. (1979). Prospect theory: An analysis of decision under risk. *Econometrica*, 47(2):263–291.

Kelley, M. (2023). On accuracy and coherence with infinite opinion sets. *Philosophy of Science*, 90(1):92–128.

Kelley, M. and Neth, S. (2023). Accuracy and infinity: a dilemma for subjective Bayesians. *Synthese*, 201(12).

Kelly, T. (2005). The epistemic significance of disagreement. In Gendler, T. S. and Hawthorne, J., editors, *Oxford Studies in Epistemology*, volume 1. Oxford University Press.

Kemeny, J. (1955). Fair bets and inductive probabilities. *The Journal of Symbolic Logic*, 20(3):263–273.

King, J. (1996). Structured propositions and sentence structure. *Journal of Philosophical Logic*, 25(5):495–521.

Kitcher, P. (1990). The division of cognitive labor. *The Journal of Philosophy*, 87(1):5–22.

Klement, K. (2025). Russell's logical atomism. *Stanford Encyclopedia of Philosophy*.

Knight, C. (2023). Reflective equilibrium. *Stanford Encyclopedia of Philosophy.*

Kocurek, A. (2021). Logic talk. *Synthese*, 199:13661–13688.

Kolodny, N. (2007). How does coherence matter? *Proceedings of the Aristotelian Society*, 107:229–263.

Konek, J. (2023). The art of learning. *Oxford Studies in Epistemology*, 7.

Kopec, M. and Titelbaum, M. (2016). The uniqueness thesis. *Philosophy Compass*, 11(4):189–200.

Korsgaard, C. (1989). Personal identity and the unity of agency: A Kantian response to Parfit. *Philosophy and Public Affairs*, 18(2):101–132.

Krantz, D., Luce, D., Suppes, P., and Tversky, A. (1971). *Foundations of Measurement*, volume 1. New York Academic Press.

Kripke, S. (1963). Semantical considerations on modal logic I normal modal propositional calculi. *Zeitschrift für mathematische Logic und Grundlagen der Mathematik*, 9:67–96.

Kripke, S. (1972). Naming and necessity. In Davidson, D. and Harman, G., editors, *Semantics of Natural Language*. Reidel.

Kyburg, H. (1961). *Probability and the Logic of Rational Belief.* Wesleyan University Press.

Lackey, J. (2008). *Learning From Words: Testimony as a Source of Knowledge.* Oxford University Press.

Landemore, H. (2011). *Democratic Reason: Politics, Collective Intelligence, and the Rule of the Many.* Princeton University Press.

Landes, J. and Williamson, J. (2013). Objective Bayesianism and the maximum entropy principle. *Entropy*, 15(9):3528–3591.

LeCun, Y., Bengio, Y., and Hinton, G. (2015). Deep learning. *Nature*, 521:436–444.

LeCun, Y., Boser, B., Denker, J. S., Henderson, D., Howard, R. E., Hubbard, W., and Jackel, L. D. (1989). Backpropagation applied to handwritten zip code recognition. *Neural Computation*, 1(4):541–551.

Lederman, H. and Mahowald, K. (2024). Are language models more like libraries or like librarians? bibliotechnism, the novel reference problem, and the attitudes of LLMs. In *Transactions of the Association for Computational Linguistics*, volume 12, pages 1087–1103.

Legg, S. and Hutter, M. (2007). Universal intelligence: A definition of machine intelligence. *Minds and Machines*, 17:391–444.

Lehrer, K. and Cohen, S. (1983). Justification, truth, and coherence. *Synthese*, pages 191–207.

Leitgeb, H. (2014). The stability theory of belief. *The Philosophical Review*, 123(2):131–171.

Leitgeb, H. and Carus, A. (2020). Rudolf carnap. *Stanford Encyclopedia of Philosophy*.

Leitgeb, H. and Pettigrew, R. (2010a). An objective justification of Bayesianism I: Measuring inaccuracy. *Philosophy of Science*, 77:201–235.

Leitgeb, H. and Pettigrew, R. (2010b). An objective justification of Bayesianism II: The consequences of minimizing inaccuracy. *Philosophy of Science*, 77:236–272.

Leslie, J. (1990). Is the end of the world nigh? *Philosophical Quarterly*, 40(158):65–72.

Levi, I. (1974). On indeterminate probabilities. *The Journal of Philosophy*, 71:391–418.

Levi, I. (1985). Imprecision and indeterminacy in probability judgment. *Philosophy of Science*, 52:390–409.

Levi, I. (1993). Rationality, prediction, and autonomous choice. *Canadian Journal of Philosophy*, 19:339–363.

Levin, J. (2013). Functionalism. *Stanford Encyclopedia of Philosophy*.

Levinstein, B. (2017). A pragmatist's guide to expected utility. *Philosophy of Science*, 84:613–638.

Levy, N. and Alfano, M. (2020). Knowledge from vice: Deeply social epistemology. *Mind*, 129(515):887–915.

Lewis, C. I. and Langford, C. H. (1932). *Symbolic Logic.* Century Company.

Lewis, D. (1970). How to define theoretical terms. *The Journal of Philosophy*, 67(13):427–446.

Lewis, D. (1976). Probabilities of conditionals and conditional probabilities. *The Philosophical Review*, 85(3):297–315.

Lewis, D. (1980). A subjectivist's guide to objective chance. In Jeffrey, R., editor, *Studies in Inductive Logic and Probability*, volume II. University of California Press.

Lewis, D. (1982). Logic for equivocators. *Noûs*, 16(3):431–441.

Lewis, D. (1986). *On the Plurality of Worlds.* Wiley-Blackwell.

Lin, H. (2013). Foundations of everyday practical reasoning. *Journal of Philosophical Logic*, 42:831–862.

Lin, H. and Kelly, K. (2012). Propositional reasoning that tracks probabilistic reasoning. *Journal of Philosophical Logic*, 41(6):957–981.

Lindley, D. (1982). Scoring rules and the inevitability of probability. *International Statistical Review*, 50(1):1–11.

List, C. and Pettit, P. (2011). *Group Agency: The Possibility, Design, and Status of Corporate Agents.* Oxford University Press.

Liu, Y. and Price, H. (2019). Heart of DARCness. *Australasian Journal of Philosophy*, 97(1):136–150.

Liu, Y. and Price, H. (2020). Ramsey and Joyce on deliberation and prediction. *Synthese*, 197:4365–4386.

Lloyd-Jones, D., Leip, E., Larson, M., D'Agostino, R., Beiser, A., Wilson, P., Wolf, P., and Levy, D. (2006). Prediction of lifetime risk for cardiovascular disease by risk factor burden at 50 years of age. *Circulation*, 113(6):791–798.

Locke, D. T. (2014). The decision-theoretic Lockean thesis. *Inquiry*, 57(1):28–54.

Locke, J. (1690). *An Essay Concerning Human Understanding.* Holt.

Loftus, E. and Pickrell, J. (1995). The formation of false memories. *Psychiatric Annals*, 25(12):720–725.

Lukits, S. (2014). The principle of maximum entropy and a problem in probability kinematics. *Synthese*, 191(7):1–23.

MacFarlane (2004). In what sense (if any) is logic normative for thought? unpublished.

Maher, P. (1995). Probabilities for new theories. *Philosophical Studies*, 77:103–115.

Mahtani, A. (2024). *The Objects of Credence*. Oxford.

Makinson, D. C. (1965). The paradox of the preface. *Analysis*, 25(6):205–207.

Malcolm, N. (1952). Knowledge and belief. *Mind*, 61(242):178–189.

Mandelkern, M. and Linzen, T. (2024). Do language models' words refer? *Computational Linguistics*, 50(3):1191–1200.

Marušić, B. (2015). *Evidence and Agence: Norms of Belief for Promising and Resolving*. Oxford.

McGee, V. (1994). Learning the impossible. In Eells, E. and Skyrms, B., editors, *Probability and Conditionals*. Cambridge University Press.

McLeod, P. and Dienes, Z. (1996). Do fielders know where to go to catch the bll or only how to get there? *Journal of Experimental Psychology: Human Perception and Performance*, 22(3):531–543.

Meacham, C. and Weisberg, J. (2011). Representation theorems and the foundations of decision theory. *Australasian Journal of Philosophy*, 89(4):641–663.

Mercier, H. and Sperber, D. (2019). *The Enigma of Reason*. Harvard University Press.

Millikan, R. (1984). *Language, Thought and Other Biological Categories*. MIT Press.

Moon, A. and Jackson, E. (2020). Credence: A belief-first approach. *Canadian Journal of Philosophy*, 50:652–669.

Moss, S. (2015a). Credal dilemmas. *Noûs*, 49(4):665–683.

Moss, S. (2015b). Time-slice epistemology and action under indeterminacy. *Oxford Studies in Epistemology.*

Moss, S. (2017). *Probabilistic Knowledge.* Oxford University Press.

Moss, S. (2018). Moral encroachment. *Proceedings of the Aristotelian Society*, 118(2):177–205.

Moss, S. (2020). Global constraints on imprecise credences: solving reflection violations, belief inertia, and other puzzles. *Philosophy and Phenomenological Research.*

Nagel, T. (1971). Brain bisection and the unity of consciousness. *Synthese*, 22:396–413.

Nash, L. (1974). *Elements of Statistical Thermodynamics.* Addison-Wesley.

Neander, K. (1991). The teleological notion of 'function'. *Australasian Journal of Philosophy*, 69(4):454–468.

Nelkin, D. (2000). The lottery paradox, knowledge, and rationality. *The Philosophical Review*, 109(3):373–409.

Nelson, E. (1977). Internal set theory: A new approach to nonstandard analysis. *Bulletin of the American Mathematical Society*, 83(6):1165–1198.

Neth, S. (2024). Better foundations for subjective probability. *Australasian Journal of Philosophy*, 103(1):1–22.

Nguyen, C. T. (2020). Echo chambers and epistemic bubbles. *Episteme*, 17(2):141–161.

Nielsen, M. (2021a). Accuracy-dominance and conditionalization. *Philosophical Studies*, 178(10):3217–3236.

Nielsen, M. (2021b). A new argument for Kolmogorov conditionalization. *Review of Symbolic Logic*, 14(4):1–16.

Nielsen, M. (2022). Regular conditional probabilities and strictly proper loss functions. *Statistics & Probability Letters*, 185.

Nielsen, M. (2023). Accuracy and probabilism in infinite domains. *Mind*, 132(526):402–427.

Nolan, D. (2021). Impossibility and impossible worlds. In Bueno, O. and Shalkowski, S., editors, *The Routledge Handbook of Modality*, pages 40–48. Routledge.

Nozick, R. (1969). Newcomb's Problem and two principles of choice. In *Essays in Honor of Carl G. Hempel*, pages 114–146. Springer.

Nozick, R. (1981). *Philosophical Explanations.* Harvard University Press.

Parfit, D. (1984). *Reasons and Persons.* Oxford University Press.

Paris, J. (1995). *The Uncertain Reasoner's Companion.* Cambridge University Press.

Paris, J. (1998). Common sense and maximum entropy. *Synthese*, 117(1):75–93.

Paul, L. (2014a). *Transformative Experience.* Oxford University Press.

Paul, S. (2014b). Diachronic incontinence is a problem in moral philosophy. *Inquiry*, 57(3):337–355.

Pettigrew, R. (2012). Accuracy, chance, and the principal principle. *Philosophical Review*, 121(2):241–275.

Pettigrew, R. (2014). Accuracy and evidence. *Dialectica*, 67(4):579–596.

Pettigrew, R. (2016a). *Accuracy and the Laws of Credence.* Oxford University Press.

Pettigrew, R. (2016b). Accuracy, risk, and the principle of indifference. *Philosophy and Phenomenological Research*, 92(1):35–59.

Pettigrew, R. (2022a). Accuracy-first epistemology without additivity. *Philosophy of Science*, 89(1):128–151.

Pettigrew, R. (2022b). *Epistemic Risk and the Demands of Rationality.* Oxford.

Pettigrew, R. and Titelbaum, M. (2014). Deference donc right. *Philosophers' Imprint*, 14(35):1–19.

Piñeros Glasscock, J. S. and Tenenbaum, S. (2023). Action. *Stanford Encyclopedia of Philosophy.*

Pinker, S. (1994). *The Language Instinct.* William Morrow and Company.

Pollock, J. (1983). A theory of direct inference. *Theory and Decision*, 15:29–95.

Popper, K. (1959). *The Logic of Scientific Discovery.* Harper & Row.

Predd, J., Seiringer, R., Lieb, E., Osherson, D., Poor, V., and Kulkarni, S. (2009). Probabilistic coherence and proper scoring rules. *IEEE Transactions on Information Theory*, 55(10):4786–4792.

Putnam, H. (1975). The meaning of 'meaning'. *Minnesota Studies in the Philosophy of Science*, 7:131–193.

Quiggin, J. (1982). A theory of anticipated utility. *Journal of Economic Behavior and Organization*, 3:323–343.

Rabinowicz, W. (2002). Does practical deliberation crowd out self-prediction. *Erkenntnis*, 57(1):91–122.

Ramsey, F. P. (1926). Truth and probability. In Braithwaite, R. B., editor, *The Foundations of Mathematics and other Logical Essays (1931).* Harcourt, Brace and Company.

Ramsey, W. (2013). Eliminative materialism. *Stanford Encyclopedia of Philosophy.*

Reed, N., McLeod, P., and Dienes, Z. (2009). Implicit knowledge and motor skill: What people who know how to catch don't know. *Conscious Cognition*, 19(1):63–76.

Rendsvig, R., Symons, J., and Wang, Y. (2023). Epistemic logic. *Stanford Encyclopedia of Philosophy.*

Rescorla, M. (2018). A Dutch book theorem and converse Dutch book theorem for Kolmogorov conditionalization. *The Review of Symbolic Logic*, 11(4):705–735.

Rinard, S. (2017). No exception for belief. *Philosophy and Phenomenological Research*, 94(1):121–143.

Robinson, A. (1966). *Non-standard analysis.* Princeton University Press.

Ross, J. and Schroeder, M. (2011). Belief, credence, and pragmatic encroachment. *Philosophy and Phenomenological Research.*

Rothschild, D. (2021). Lockean beliefs, Dutch books, and scoring systems. *Erkenntnis*, 88(5):1979–1995.

Roush, S. (2005). *Tracking Truth: Knowledge, Evidence, and Science.* Oxford.

Royall, R. (1997). *Statistical Evidence: A Likelihood Paradigm.* Chapman and Hall.

Rumelhart, D., Hinton, G., and Williams, R. (1986). Learning representations by back-propagating errors. *Nature*, 323(9):533–536.

Ryan, S. (1996). The epistemic virtues of consistency. *Synthese*, 109:121–141.

Samuelson, P. (1938). A note on the pure theory of consumer's behavior. *Economica*, 5(17):61–71.

Samuelson, P. (1948). Consumption theory in terms of revealed preference. *Economica*, 15(60):243–253.

Savage, L. J. (1954). *The Foundations of Statistics.* Dover.

Schervish, M. J. (1989). A general method for comparing probability assessors. *Annals of Statistics*, 17:1856–1879.

Schervish, M. J., Seidenfeld, T., and Kadane, J. B. (2009). Proper scoring rules, dominated forecasts, and coherence. *Decision Analysis*, 6(4):202–221.

Schoenfield, M. (2012). Chilling out on epistemic rationality: A defense of imprecise credences (and other imprecise doxastic attitudes). *Philosophical Studies*, 158(2):197–219.

Schoenfield, M. (2014). Permission to believe: Why permissivism is true and what it tells us about irrelevant influences on belief. *Noûs*, 48(2):193–218.

Schoenfield, M. (2016). The accuracy and rationality of imprecise credences. *Noûs.*

Schoenfield, M. (2017). Conditionalization does not maximize expected accuracy. *Mind*, 126(504):1155–1187.

Schroeter, L. (2017). Two-dimensional semantics. *Stanford Encyclopedia of Philosophy.*

Schultheis, G. (2023). Accurate updating. *Philosophy of Science.*

Schwitzgebel, E. (2001). In-between believing. *Philosophical Quarterly*, 51:76–82.

Segall, M. H., Campbell, D. T., and Herskovits, M. J. (1963). Cultural differences in the perception of geometric illusions. *Science*, 139:769–771.

Seidenfeld, T., Kadane, J. B., Schervish, M. J., and Stern, R. B. (2024). Finite additivity, complete additivity, and the comparative principle. *Erkenntnis*, 90(5):1945–1968 (2024).

Sellars, W. (1956). Empiricism and the philosophy of mind. In Feigl, H. and Scriven, M., editors, *Minnesota Studies in the Philosophy of Science, Volume I: The Foundations of Science and the Concepts of Psychology and Psychoanalysis*, pages 253–329. University of Minnesota Press.

Sharadin, N. (2018). Epistemic consequentialism: Haters gonna hate. In Kyriacou, C. and McKenna, R., editors, *Metaepistemology: Realism & Antirealism*, pages 121–143. Palgrave Macmillan.

Shear, T. and Fitelson, B. (2019). Two approaches to belief revision. *Erkenntnis*, 84(3):487–518.

Silver, K. (2022). Group action without group minds. *Philosophy and Phenomenological Research*, 104(2):321–342.

Simon, H. (1990). Invariants of human behavior. *Annual Review of Psychology*, 41:1–19.

Singer, D. (2019). Permissible epistemic trade-offs. *Australasian Journal of Philosophy*, 97(2):281–293.

Singer, D. and Aronowitz, S. (2022). What epistemic reasons are for: Against the belief-sandwich distinction. In Dunaway, B. and Plunkett, D., editors, *Meaning, Decision, and Norms: Themes from the Work of Allan Gibbard.* Maize Books.

Skipper, M. and Bjerring, J. C. (2019). Bayesianism for the average Joe. *Erkenntnis*, 7(1):93–115.

Skyrms, B. (1984). *Pragmatics and Empiricism.* Yale University Press.

Skyrms, B. (1985). Maximum entropy inference as a special case of conditionalization. *Synthese*, 63(1):55–74.

Skyrms, B. (1987). Dynamic coherence and probability kinematics. *Philosophy of Science*, 54:1–20.

Skyrms, B. (1992). Coherence, probability and induction. *Philosophical Issues*, 2:215–226.

Soares, N. and Levinstein, B. (2019). Cheating death in Damascus. *The Journal of Philosophy.*

Sosa, E. (1999). How to defeat opposition to Moore. *Philosophical Perspectives*, 33(141-153).

Soysal, Z. (2024). An algorithmic impossible-worlds model of belief and knowledge. *The Review of Symbolic Logic*, 17(2):586–610.

Spelke, E. and Kinzler, K. (2007). Core knowledge. *Developmental Science*, 10(1):89–96.

Staffel, J. (2017). Should I pretend I'm perfect? *Res Philosophica*, 94(2):301–324.

Stalnaker, R. (1984). *Inquiry.* MIT Press.

Steinberger, F. (2017). Consequence and normative guidance. *Philosophy and Phenomenological Research*, 98(2):306–328.

Steinberger, F. (2019). Three ways in which logic might be normative. *Journal of Philosophy*, 116(1):5–31.

Strohminger, N. and Nichols, S. (2015). Neurodegeneration and identity. *Psychological Science*, 26(9):1469–1479.

Sturgeon, S. (2008). Reason and the grain of belief. *Noûs*, 42:139–165.

Swinburne, R. (1971). The paradoxes of confirmation a survey. *American Philosophical Quarterly*, 8(4):318–330.

Talbott, W. J. (1991). Two principles of Bayesian epistemology. *Philosophical Studies*, 62(2):135–150.

Tarsney, C. (2020). Exceeding expectations: Stochastic dominance as a general decision theory. GPI Working Paper No. 3-2020.

Teller, P. (1973). Conditionalization and observation. *Synthese*, 26:218–258.

Terven, J., Cordova-Esparza, D.-M., Romero-González, J.-A., and Ramírez-Pedraza, A., and Chávez-Urbiola, E. A. (2025). A comprehensive survey of loss functions and metrics in deep learning. *Artificial Intelligence Review*, 58(195).

Thorstad, D. (2024). Why bounded rationality (in epistemology)? *Philosophy and Phenomenological Research*, 108(2):396–413.

Titelbaum, M. (2012). *Quitting Certainties.* Oxford University Press.

Titelbaum, M. (2022a). *Fundamentals of Bayesian Epistemology 1: Introducing Credences.* Oxford University Press.

Titelbaum, M. (2022b). *Fundamentals of Bayesian Epistemology 2: Arguments, Challenges, Alternatives.* Oxford University Press.

Todd, P. M. and Brighton, H. (2015). Building the theory of ecological rationality. *Minds and Machines*, 26:9–30.

Tollefsen, D. (2015). *Groups as Agents.* Polity Press.

Turing, A. (1950). Computing machinery and intelligence. *Mind*, 49(433-460).

van Fraassen, B. (1980). *The Scientific Image.* Clarendon Press.

van Fraassen, B. (1984). Belief and the will. *The Journal of Philosophy*, 81(5):235–256.

van Fraassen, B. (1995). Belief and the problem of Ulysses and the Sirens. *Philosophical Studies*, 77:7–37.

van Fraassen, B. (1999). Conditionalization, a new argument for. *Topoi*, 18:93–96.

Van Leeuwen, N. and Lombrozo, T. (2023). The puzzle of belief. *Cognitive Science*, 47.

Vassend, O. (2015). Confirmation measures and sensitivity. *Philosophy of Science*, 82(5):892–904.

Vassend, O. (2018). Goals and the informativeness of prior probabilities. *Erkenntnis*, 83(4):647–670.

Vaswani, A., Shazeer, N., Parmar, N., Uszkoreit, J., Jones, Llion and Gomez, A., Kaiser, L., and Polosukhin, I. (2017). Attention is all you need. In *31st Conference on Neural Information Processing Systems*.

Vermaire, M. (2025). Judgment's aimless heart. *Noûs*, 59(1):3–21.

Vesga, A., Van Leeuwen, N., and Lombrozo, T. (2025). Evidence for multiple kinds of belief in theory of mind. *Journal of Experimental Psychology: General*.

Vineberg, S. (2022). Dutch book arguments. *Stanford Encyclopedia of Philosophy*.

von Wright, G. H. (1951). *An Essay in Modal Logic*. North-Holland.

Wagner, C. (2002). Probability kinematics and commutativity. *Philosophy of Science*, 69:266–278.

Walley, P. (1991). *Statistical Reasoning with Imprecise Probabilities*. Chapman and Hall.

Wang, Q., Ma, Y., Zhao, K., and Tian, Y. (2022). A comprehensive survey of loss functions in machine learning. *Annals of Data Science*, 9:187–212.

Weatherall, J. O., O'Connor, C., and Bruner, J. (2020). How to beat science and influence people: Policymakers and propaganda in epistemic networks. *The British Journal for the Philosophy of Science*, 71(4):1157–1186.

Weatherson, B. (2005). Can we do without pragmatic encroachment? *Philosophical Perspectives*, 19(1):417–443.

White, R. (2009). Evidential symmetry and mushy credence. *Oxford Studies in Epistemology*.

Williamson, T. (2002). *Knowledge and its Limits*. Oxford.

Williamson, T. (2011). Improbable knowing. In Dougherty, T., editor, *Evidentialism and its Discontents*. Oxford University Press.

Yalcin, S. (2012). Bayesian expressivism. *Proceedings of the Aristotelian Society*, 112(2):123–60.

Yudkowsky, E. (2010). Timeless decision theory. Technical report, The Singularity Institute.

Yudkowsky, E. and Soares, N. (2018). Functional decision theory: a new theory of instrumental rationality. Technical report, Machine Intelligence Research Institute.

Zollman, K. (2007). The communication structure of epistemic communities. *Philosophy of Science*, 74(5):574–587.

Zynda, L. (2000). Representation theorems and realism about degrees of belief. *Philosophy of Science*, 67:45–69.

Index